少廢話！英文E-mail別拖泥帶水，講重點就好！
快點抄！複製→貼上→寄出→達成目的超高效！

學習有捷徑
夢想最接近

Point 1

寫作要點
Key Points

本書首先以11大主題，共70個小情境將英文E-mail分類後，每篇小情境的英文書信又以3大段落個別解析，共分為：開頭（Introduction）、本體（Body）與結尾（Closing）。藉由清楚明確的分段，讓你可以依照各篇一開始所提示的段落主旨找到大方向，再以中文擬出信件大綱，從此不必擔心自己在撰寫英文信件時，一不留神就寫得落落長，或者邏輯鬆散、文不對題。

01 詢問職位空缺

A Job Inquiry Letter

寫作要點 Key Points:

Step 1: 清楚表示來信目的是為求職。可以的話，說明為何該公司吸引你求職。
Step 2: 簡短介紹自己的專業領域及感興趣的職務內容。
Step 3: 隨信附上履歷，並強調希望能夠得到對方回覆。

實際E-mail範例

| 寫信 |▼ | 刪除 | 回覆 |▼ | | 寄件者： Ginny |

Dear Sir/Madam,

Introduction

I am writing this letter to inquire about job vacancies at your company. For th past few years I have followed your company through news and your officia website, and I am very impressed by your dedication to the news media.

Body

I am currently a news journalist and have been engaged in news work for five years since I graduated from National Cheng Chi University.
I am able to work with general current affairs and I can also specialize in certain issues, such as political or business issues.

Closing

My attached CV provides additional details about my professional background.
I would be delighted to discuss any possible job vacancies with you at your convenience.

I am looking forward to hearing from you in the near future.

Best regards,
Signature of Sender
Sender's Name Printed

016

Point 2

實際E-mail範例

掌握信件各段落的主旨與大綱後，便可以參考各篇中收錄的商務實境E-mail範例，實際了解適用於開頭、本體與結尾的各種實用英文句型、專業字彙，以及如何以最精簡的方式組織出條理分明的英文E-mail，深化各種商務情境下英文信件的寫作概念。

Point 3　各段落超實用句型

　　一再重覆使用相同的句型與詞彙，是寫作各式商務英文信件的大忌，即使是段落條理清楚、主旨分明，也會因為一再重覆的句型與單調乏味的用字，令閱讀信件的國外客戶感到不耐，甚至看輕自己。此時不妨參考各段落精選的各種實用句型，讓你撰寫英文信件時，不再一種句型用到底，寫出來的英文E-mail自然流暢精準，令人眼睛一亮！

---

e-mail中譯　　寫信▼ | 刪除 | 回覆▼ | 寄件

好：

題
寫這封信的目的，是要詢問貴公司的工作職缺。過去幾年我一直在新聞及您們
網上追蹤貴公司的消息，並且對於貴公司在新聞媒體領域的貢獻有深刻的印

前是個新聞記者，自政治大學畢業後就投身新聞工作長達五年。我能負責一
時事報導，亦能負責特定議題，諸如政治議題或商業議題等。

於我更多的專業背景，均詳細記錄在隨信附上的履歷表中。我將樂意在任何
便的時間與您討論可能的工作機會。期望近日能得到您的回覆。

安康，
／署名

段落超實用句型

：畫底線部分的單字可按照個人情況自行替換

troduction

urpose of this E-mail is to enquire for any vacancies in your
any.
此封郵件的目的是為詢問貴公司目前是否有任何工作職缺。

ld like to inquire whether you have any openings in your
e department for which I might apply.
問貴公司的財務部門是否有任何我可以應徵的職缺。

you for taking the time to review my resume. I have just
ted from Cheng Kung University, and I am currently looking
time employment in Tainan City.
冗閱讀我的簡歷。我甫自成功大學畢業，目前正在大台南地區尋覓全職

求職求才篇　Part 1

❷ I will appreciate an opportunity to speak with you regarding how my skills and experiences could fill your needs.
若有機會與您談談如何使我的技能及經驗符合您的職務需求，我將不勝感激。

❸ Should you have any questions regarding my qualifications, please feel free to contact me by my cellphone 0912345678 or E-mail me at myname@gmail.com.
若您對我的求職資格有任何問題，歡迎您撥打我的手機 0912345678，或來信 myname@gmail.com 與我聯絡。

❹ Attached to this E-mail you will find my CV for consideration for the post of Quality Control Manager.
隨此郵件附上我的履歷，以供您作為由我出任品管經理一職之參考。

❺ In case there are not any suitable openings at present, I will appreciated it very much if you keep my CV on file for future possibilities.
萬一目前沒有適合的空缺，若您能將我的履歷留檔作為日後求才參考，我將感激不盡。

✎ 不可不知的實用E-mail字彙

- CV= curriculum vitae 履歷；簡歷
- certificate 證明書
- professional license 專業證照；執業證照
- practice license 執業執照
- accountant license 會計師執照
- lawyer license 律師執照
- pharmacist license 藥師執照
- nutritionist license 營養師執照
- teacher license 教師執照
- Test of English for International Communication（TOEIC
  語能力認證

Point 4　不可不知的實用E-mail字彙

　　全書70篇商務實境E-mail，篇篇皆收錄外商職場情境中的重點專業詞彙、實用單字與高頻率片語，讓你沒有英文字典也不怕拼錯字，直接隨抄隨用，用字遣詞不再單調無味，在國外客戶心中留下絕佳印象，在撰寫信件時更能利用情境順便完整學習相關的英文字彙！

　　我曾任職外商公司國際行銷業務多年，在工作期間，我發現許多國外業務同事最擔心的就是不知道該如何撰寫得體的英文E-mail，適切地向外國客戶表達自己的主題及重點。另外，我也觀察到某些國外業務，因為本身英文能力的限制，而以最粗淺的英語詞彙撰寫商務信件，最後往往在業務往來上吃了悶虧，甚至遭到國外客戶的輕視和刁難。有鑑於此，我深深覺得若能掌握商務英文信件的寫作要領，在職場上就能如虎添翼、無往不利。

　　在跨足國際行銷領域之前，我曾經投身於英語學習與英語教學領域10多年。當開始從事外商公司行銷業務後，為了寫出讓外國客戶印象深刻的英文信件，我便在短時間內以過去英語學習與教學的經驗，找到撰寫商務英文信件的關鍵要領：「英文E-mail不用落落長，精準得體最重要」，也就是在一封不足100字的英文E-mail中，以專業詞彙與精簡短句精確地傳達自己的想法，讓外國客戶對你印象深刻。

　　我希望能以同為「國外業務過來人」的經驗，撰寫一本包含各種情境的商務英文E-mail工具書，將我多年來商務英文信件的撰寫經驗及寫作要領與大家分享。而本書就是我歸納外商職場上70種實境的商務信件，囊括介紹提案、通知邀約、報價訂單、出貨付款等的商務英文書信精華大全，希望讀者能一本在手，面對商務英文E-mail時輕鬆過關斬將，達到好業績！

Wenny Tsai

Contents 目·錄

使用説明／002　　　商務英文E-mail寫作的7C原則／008
作者序／004　　　　商務英文E-mail開頭&結尾實用短句各20則／011

 Part1：求職求才篇 ／014

01 詢問職位空缺／016　　　**04** 面試後感謝信／031
02 撰寫應徵履歷信／021　　**05** 邀請面試告知信／035
03 詢問面試結果／026　　　**06** 職位錄取告知信／039

 Part2：各式介紹篇 ／044

01 向同事自我介紹／046　　**04** 向廠商介紹產品／服務／058
02 向廠商自我介紹／050　　**05** 向廠商介紹新同事／團隊／062
03 向廠商介紹公司／054

 Part3：簡易提案篇 ／066

01 向廠商提出方案／068　　**04** 更改方案內容／080
02 贊成廠商的方案／072　　**05** 協調方案內容，尋求共識／084
03 拒絕廠商的方案／076

 Part4：活動邀請篇 ／088

01 詢問出席聚會意願／090　　**05** 告知活動取消／106
02 詢問出席會議意願／094　　**06** 答應他人邀請／110
03 告知活動時間與流程／098　　**07** 婉拒他人邀請／114
04 更改活動時間或流程／102

Part5：訂單交涉篇 ／118

01 要求廠商估價 ／120
02 向廠商提供估價 ／124
03 向廠商提出議價 ／128
04 確認報價並下單 ／132
05 拒絕廠商的報價 ／136
06 變更訂單或取消訂單 ／140

Part6：詢問與請求篇 ／144

01 詢問公司資訊 ／146
02 詢問訂單進度 ／150
03 詢問款項進度 ／154
04 請求對方介紹產品／服務 ／158
05 請求退費 ／162
06 請求提供樣品 ／166
07 請求提供建議 ／170
08 請求延後交貨 ／174
09 請求延後付款 ／178

Part7：各式通知篇 ／182

01 出貨通知 ／184
02 缺貨通知 ／188
03 休假通知 ／192
04 人事異動通知 ／196
05 地址、電話變更通知 ／200
06 會議記錄通知 ／204

Part8：進度催促篇 ／208

01 催促出貨 ／210
02 催促返還借出之資料、樣品 ／214
03 催促提供發票 ／218
04 催促支付款項 ／222
05 催促返還合約 ／226

Part9：感謝與道歉篇 /230

01 臨時缺席／與會取消道歉 /232
02 延遲出貨道歉 /236
03 貨物瑕疵、毀損道歉 /240
04 貨款滯納道歉 /244
05 貨物錯誤道歉 /248
06 意外違反合約道歉 /252
07 感謝活動款待 /256
08 感謝下訂／合作 /260
09 感謝協助 /264

Part10：誠心祝賀篇 /268

01 祝賀職位升遷 /270
02 祝賀公司喬遷 /274
03 祝賀各佳節 /278
04 祝賀開業、公司擴展 /282

Part11：其他重要信件 /286

01 請求聯絡與回覆 /288
02 向主管或人事單位請假 /292
03 回覆抱怨與客訴 /296
04 提出離職要求 /300
05 推薦合適人選 /304
06 哀悼與慰問 /308
07 對外澄清誤會 /312
08 公司內部物品申請 /316

商務英文E-mail寫作的7C原則

在職場上利用 E-mail 傳達訊息時，其實並不需要詞藻華麗、文句優美；真正需要做的就是用「簡單、精準的語句，清晰、準確且完整地表達自己的意思」，讓對方可以非常清楚地瞭解您想傳達的訊息。那麼何謂商務英文 E-mail 寫作務必掌握的「7C 原則」？請看以下針對「7C」的詳盡說明。

1. Conciseness 簡潔

優秀的英文 E-mail 應該行文流暢、言簡意賅，並且避免過度冗長。也就是說，寫信者必須使用簡短精確的文字表達重要資訊。

請比較下面2個句子：

❶ Our company's operations for the preceding accounting period terminated with a substantial deficit.（本公司在去年的帳務上呈現巨額的赤字。）→（冗長）

❷ Our company had a substantial deficit last year.（本公司在去年的帳務上呈現巨額的赤字。）→（簡潔）

2. Concreteness 具體

英文 E-mail 在書寫過程中，當涉及一些情況的描述，例如：時間、地點、價格、貨物品質等，寫信者應盡可能以簡短的文句作具體描述，如此便可以使溝通的內容更加清楚、明確，並且有助於加快事務的進程。

請比較下面2個句子：

❶ These brakes can stop a car within a short distance.（這項煞車系統可以在短距離內使一台車停下來。）→（尚可）

❷ These Goodson power brakes can stop a 2-ton car within24 feet.（古森煞車系統可以在24英呎內使一台兩噸重的車停下。）→（更具體）

3. Clearness 清晰

「清晰」是英文書信寫作最重要的原則。一封含糊不清、詞不達意的書信，很容易引起誤會與意見分歧，甚至會造成業務上的損失。寫作上應選擇精確的詞彙，

並且使用句型結構簡明的短句。一般來説，商業書信的撰寫者必須以正確的句子，將自己的主張清晰地表達出來，以便對方準確理解。

請比較下面2個句子：

❶ We can supply 70 tons of the item only.（我們只能提供此項商品七十噸。）
→（著重於能夠提供的「商品品項」）

❷ We can supply only 70 tons of the item.（我們只能提供七十噸的此項商品。）→（著重於能夠提供的「數量」）

4. Courtesy 禮貌

即使以 E-mail 聯繫對方，同樣需要注意禮儀。而在寫信過程中，也要避免太過直接、會傷害對方感受的表達方式。措辭上應選用委婉的詞語，像 would, could, may, please等。當然這並不代表要一味低三下四地討好對方，只要在措辭上避免冒犯對方即可。

請比較下面2個句子：

❶ We are sorry that you misunderstood me.（我們很遺憾您對我們的誤解。）
→（「責備對方」誤解己方）

❷ We are sorry that we didn't make ourselves clear.（我們很抱歉沒把我們的論點表達清楚。）→（「就自己的疏失」向對方道歉）

5. Consideration 體貼

在英文 E-mail 寫作過程中，撰寫者應設身處地替對方著想，尊重對方的風俗習慣，即採取所謂的「You-Attitude」（對方的立場），並盡可能避免使用「I-Attitude」或「We-Attitude」（我方的立場）。另外，還應該考慮到收信者的文化習慣、性別等各方面的因素。

請比較下面2個句子：

❶ I received your letter of June 23 rd, 2017 this morning.（我於2017年6月23日上午收到了您的信件。）→（我方的立場）

❷ Your letter of June 23rd, 2017 arrived this morning.（您的信件已於2017年6月23日上午寄達。）→（對方的立場）

6. **C**orrectness 正確

在英文 E-mail 中，除了避免文法、拼寫及標點符號錯誤外，其中所引用的資料、數據等也應準確無誤。尤其在商務英語中提到具體日期、資金、數量等內容時更要準確表達以免發生歧義與糾紛。

請比較下面2個句子：

❶ This contract will come into effect from Oct.1st.（這份合約將於10月1日起生效。）→（資訊不精準）

❷ This contract will come into effect from and including October 1 st, 2016.（這份合約將於 2016年10月1日當日開始生效。）→（資訊精確）

7. **C**ompleteness 完整

在寫作商務英文 E-mail時，為使信件具備完整性，信函中應包括所有必要的資訊。例如下列的通知內容，雖然才寥寥幾句就包含了所有的重要資訊。

請參考下列內容：

Notice:

All the staffs of Accounts Department are requested to be ready to attend the meeting in the conference room on Tuesday, at 3:00 p.m., Jun. 6, 2016, to discuss the financial statement of last year.

Accounts Department

（通知：所有會計部同仁都必須參加2016年6月6日星期二下午3點於會議室舉行的會議，在會議上我們將會討論去年的財務報告。 會計部）

商務英文E-mail
開頭&結尾實用短句各20則

　　厭倦了自己在商務英文信件上一再重覆的開頭與結尾了嗎？不妨參考以下開頭、結尾短句各20則，隨抄隨用，讓自己的商務英文書信更有變化！

I. 商務英文E-mail開頭實用短句：

說明：畫底線部份的單字可按照個人情況自行替換

1-1. 為收到對方的來信表達感謝／欣喜

01. Thank you for your letter dated June 20 th, 2017.

02. Your letter that arrived today gave me great comfort.

03. What a treat to receive your kind letter of May 5th!

04. It is always a thrill to receive your e-mail/letter.

05. With great delight, I learned from your letter of this Sunday that...

06. I was so glad to receive your letter of March 23rd.

07. I am very obliged to you for your warm congratulations.

08. We acknowledge with thanks the receipt of your letter dated Feb 5th.

1-2. 為無法即時回覆對方表達歉意

09. I regret being unable to reply to your letter earlier due to pressure of work.

10. I hope that you will excuse me for this late reply to your kind letter.

11. I must apologize for not being able to reply to your kind letter.

1-3. 為討論對方來信中的內容

12. I am writing to ask about the conference to be held in New York next week.

13. We learn from your e-mail that you are interested in our products.

14. In reply to your letter of May 16th,...

1-4. 為通知對方己方新資訊／好消息

15. We are pleased (glad) to inform you that...

16. We are pleased to send you our catalog.

17. I am very much pleased to inform you that my visit to your company has been approved.

18. I wish to apply for the <u>teaching position</u> you are offering.

1-5. 為祝賀對方

19. I am very excited and delighted at your good news.

1-6. 為感謝對方

20. First of all, I must thank you for your kind assistance and attention to me.

 II. 商務英文E-mail結尾實用用語：

說明：畫底線部份的單字可按照個人情況自行替換

1-1. 表達希望盡快得到對方正面回覆

01. Your early reply will be highly appreciated.

02. Hoping to receive your early reply.

03. Looking forward to your early reply.

04. Awaiting your good news.

1-2. 表示對下次會面的期待

05. I look forward to our next meeting in <u>Los Angeles</u>.

06. Can't wait to meet you in <u>the Conference</u> <u>next year</u>.

1-3. 祝福對方／對方家人

07. I wish you every success in the coming year.

08. With best regards to your family.

09. I hope everything will be well with you.

10. Best wishes for all of you.

1-5. 感謝對方

11. The help you gave me is sincerely valued.

12. Your attention is appreciated.

13. Thank you very much for your consideration.

14. I appreciate your immediate reply. Thank you again!

15. Please accept my sincere thanks for your kind attention to this matter.

1-5. 歡迎對方進一步詢問或諮詢

16. Please feel free to contact me if further information is required.

17. If you need any assistance, I am available any time.

18. Please let me know if you require further information.

19. I am always glad to be of serving to you.

20. Please do not hesitate to contact me if you...

Part1
求職求才篇

01 詢問職位空缺

A Job Inquiry Letter

Step 1: 清楚表示來信目的是為求職。可以的話，說明為何該公司吸引你求職。
Step 2: 簡短介紹自己的專業領域及感興趣的職務內容。
Step 3: 隨信附上簡歷，並強調希望能夠得到對方回覆。

 實際E-mail範例

| 寫信 ▼ | 刪除 | 回覆 ▼ | 寄件者： Ginny |

Dear Sir/Madam,

Introduction

I am writing this letter to inquire about job vacancies at your company. For the past few years I have followed your company through news and your official website, and I am very impressed by your dedication to the news media.

Body

I am currently a news journalist and have been engaged in news work for five years since I graduated from National Cheng Chi University.

I am able to work with general current affairs and I can also specialize in certain issues, such as political or business issues.

Closing

My attached CV provides additional details about my professional background. I would be delighted to discuss any possible job vacancies with you at your convenience.

I am looking forward to hearing from you in the near future.

Best regards,
Signature of Sender
Sender's Name Printed

E-mail中譯

寫信｜▼　刪除　回覆｜▼　寄件者：Ginny

您好：

開頭

我寫這封信的目的，是要詢問貴公司的工作職缺。過去幾年我一直在新聞及您們的官網上追蹤貴公司的消息，並且對於貴公司在新聞媒體領域的貢獻有深刻的印象。

本體

我目前是個新聞記者，自政治大學畢業後就投身新聞工作長達五年。我能負責一般的時事報導，亦能負責特定議題，諸如政治議題或商業議題等。

結尾

有關於我更多的專業背景，均詳細記錄在隨信附上的履歷表中。我將樂意在任何您方便的時間與您討論可能的工作機會。期望近日能得到您的回覆。

敬祝安康，
簽名檔／署名

各段落超實用句型

說明：畫底線部分的單字可按照個人情況自行替換

開頭 Introduction

❶ **The purpose of this E-mail is to enquire for any vacancies in your company.**

我撰寫此封郵件的目的是為詢問貴公司目前是否有任何工作職缺。

❷ **I would like to inquire whether you have any openings in your finance department for which I might apply.**

我想詢問貴公司的財務部門是否有任何我可以應徵的職缺。

❸ **Thank you for taking the time to review my resume. I have just graduated from Cheng Kung University, and I am currently looking for full-time employment in Tainan City.**

感謝您撥冗閱讀我的簡歷。我甫自成功大學畢業，目前正在大台南地區尋覓全職工作。

❹ I am interested in applying for a position in your <u>research and development department</u>. I have heard that <u>USC</u> is a wonderful company to work for, and I hope that I can be considered for the team.

我有興趣應徵貴公司<u>研發部門</u>的職務。久聞 <u>USC</u> 是間很棒的公司,我希望貴公司能考慮讓我加入團隊。

❺ I am submitting my resume to you for consideration for the <u>office assistant</u> position in your organization.

謹以此信提交個人履歷,以供您考慮提供我貴公司<u>辦公室助理</u>一職。

本體 Body

❶ I have recently finished a diploma in <u>Business Administration</u> and have gained a <u>professional license in accounting</u>.

我已於近日取得<u>商業管理</u>的學位文憑,並拿到<u>職業會計師執照</u>。

❷ I am a successful <u>sales executive</u> with over <u>ten years</u> experience in <u>sales</u>.

我是個在<u>業務領域</u>擁有超過<u>十年</u>經驗的成功<u>業務經理人</u>。

❸ I would like to pursue a career in <u>product marketing</u>. I am interested in any entry-level position and I am willing to take any training that is required.

我想應徵<u>產品行銷</u>的職位。我對任何初階職務都有興趣,並願意接受必要的訓練。

❹ The nature of my <u>master's course</u> and my <u>two-year</u> working experience has prepared me well for this position.

<u>研究所課程</u>所學以及<u>兩年</u>的工作經驗已讓我充分準備好勝任此職務。

❺ I have over <u>five years</u> of experience in <u>public relations</u> and I am proficient in English, <u>German</u> and <u>Spanish</u>. I am willing to travel for work.

我有超過<u>五年</u>的公關經驗,精通英語、<u>德語</u>及<u>西班牙語</u>,並且願意出差工作。

結尾 Closing

❶ I have attached my CV with this inquiry letter. I am available for interview at any time.

我已經隨此詢問信函附上我的履歷表。我方便於任何時間前往面試。

❷ I will appreciate an opportunity to speak with you regarding how my skills and experiences could fill your needs.

若能有機會與您談談如何使我的技能及經驗符合您的職務需求，我將不勝感激。

❸ Should you have any questions regarding my qualifications, please feel free to contact me by my cellphone 0912345678 or E-mail me at myname@gmail.com.

若您對我的求職資格有任何問題，歡迎您撥打我的手機 0912345678，或來信 myname@gmail.com 與我聯絡。

❹ Attached to this E-mail you will find my CV for consideration for the post of Quality Control Manager.

隨此郵件附上我的履歷，以供您作為由我出任品管部經理一職之參考。

❺ In case there are not any suitable openings at present, I will appreciated it very much if you keep my CV on file for future possibilities.

萬一目前沒有適合的空缺，若您能將我的履歷留檔作為日後求才參考，我將感激不盡。

 不可不知的實用E-mail字彙

- **CV= curriculum vitae**　履歷；簡歷
- **certificate**　證明書
- **professional license**　專業證照；執業證照
- **practice license**　執業執照
- **accountant license**　會計師執照
- **lawyer license**　律師執照
- **pharmacist license**　藥師執照
- **nutritionist license**　營養師執照
- **teacher license**　教師執照
- **Test of English for International Communication（TOEIC）**　多益國際英語能力認證

- **Test of English as a Foreign Language（TOEFL）** 托福國際英語能力認證
- **International English Language Testing System （IELTS）** 雅思國際英語能力認證
- **Certificate of General English Proficiency Test (LTTC GEPT), Elementary Level** 全民英檢初級合格證書
- **Certificate of General English Proficiency Test (LTTC GEPT), Intermediate Level** 全民英檢中級合格證書
- **Certificate of General English Proficiency Test (LTTC GEPT), High-Intermediate Level** 全民英檢中高級合格證書
- **Certificate of General English Proficiency Test (LTTC GEPT), Advanced Level** 全民英檢高級合格證書
- **International Project Management Certificate of Trainer** IPMA D 級專案管理證書
- **Securities Specialist** 證券商業務員證照
- **Securities Investment Trust and Consulting Professional** 投信投顧業務員證照
- **Securities Investment Analyst** 證券分析師證照
- **Futures Specialist** 期貨業務員證照
- **Bond Specialist** 債券人員證照
- **Professional Program design Engineer** TQC 專業程式設計工程師
- **Professional Multimedia Webpage Designer** TQC 專業多媒體網頁設計工程師
- **YMCA Personal Training Instructor** YMCA 國際體適能教練證照
- **Linux System Administration (Professional)** Linux 系統管理（專業級）證照
- **Linux Network Management (Professional)** Linux 網路管理（專業級）證照
- **Chinese Institute of Industrial Engineers for professional attainment in Industrial Engineering** 中國工業工程師證照
- **Certificate for Professional Test on Trust Businesses** 信託業務專業測驗合格證明書
- **Test of English for International Communication** 多益國際英語能力認證
- **technician certificate** 技師證書

02 撰寫應徵履歷信

A Cover Letter

實際E-mail範例

寫信｜▼　刪除　回覆｜▼　寄件者： Ginny

Dear Sir/Madam,

Introduction

Please find the attachment for my CV in application for the post advertised on Manpower Agency on May 15th.

Body

As you can see in my resume, I have had extensive work experience in retail sales and service industries, where I gained skills and the ability to work with different types of customers and deal with complaints. In addition to these, I am also proficient in meeting sales targets. Also, I am very flexible and adapt well to different situations and new environments. I am confident that I could fit into your team without difficulty.

Closing

Please review my attached CV for a more detailed account of my professional skills and background.

I am available to start working immediately and I look forward to speaking with you at your convenience.

Thank you for your consideration.

Best regards,
Signature of Sender
Sender's Name Printed

E-mail中譯

| 寫信 ▼ | 刪除 | 回覆 ▼ | 寄件者： Ginny |

您好：

開頭

為了應徵貴公司於 5 月 15 日刊登於人力銀行的徵才啟示，隨信附上我的個人履歷。

本體

如您在我的履歷中所見，我在零售業務以及服務業有長足的工作經驗，使我能夠與不同類型的顧客共事，並具有處理抱怨投訴的能力。除此之外，我具備達到業務目標的能力。再加上我個性靈活有彈性，能夠適應不同的情況與新環境。我有自信能很快就融入您的團隊。

結尾

有關於我更詳細的專業技能及背景，請參閱隨信附上的個人履歷。

一經錄取我便可以立即開始上班，很期待能在您方便的時間與您面談。

感謝您的考量。

敬祝安康，
簽名檔／署名

各段落超實用句型

說明：畫底線部分的單字可按照個人情況自行替換

開頭 Introduction

❶ I learned from the advertisement you posted on the 104 Manpower Agency on September 24th that you currently have a vacancy for Sales Manager.

我從您於 9 月 24 日刊登在 104 人力銀行的廣告得知貴公司目前有個業務經理的職缺。

❷ I saw your advertisement for an office secretary and would like to express my interest.

我看到您徵求一名辦公室秘書的廣告，於是想要表達我對此職務的興趣。

❸ I am writing this letter to express my interest in the position of Sales Assistant.

謹以此信表達我對業務助理一職的興趣。

❹ This letter is an expression of my keen interest in the vacancy for a Maintenance Engineer which you advertised on 1234 Manpower Website on November 12th.

僅以此信表達我對您於 11 月 12 日刊登在 1234 人力網站上，徵求維修工程師一職的強烈興趣。

❺ With this cover letter, I would like to let you know that I am interested in offering my service as a Purchasing Assistant in your company.

藉由此應徵信函，我想讓您知道我對貴公司的採購助理一職非常有興趣。

❻ I am highly interested in the position of Senior Accountant your company is offering, and would be grateful for an opportunity to have an interview with you.

我對您公司高階會計的職位非常感興趣，若有機會接受面試，我將感激不盡。

本體 Body

❶ I have three years of working experience in magazine editing.

我在雜誌編輯方面有三年的工作經驗。

❷ My last job was with Jacob's Publishing House as a chief editor in charge of leading a team in the Editorial Department.

我的上一份工作是在雅各出版社擔任主編，負責在編輯部門領導一個團隊。

❸ I had worked as an intern engineer with TSMC for two years, which has given me professional and practical experience in the field of Process Integration.

我曾在台積電擔任實習工程師兩年，使我累積在製程整合領域的專業能力以及實務經驗。

❹ My sufficient experience in material editing would allow me to make an immediate contribution to your organization.

我在教材編輯方面的充足經驗將使我能對貴公司做出立即的貢獻。

❺ For the past two years I have worked for a well-known trading company where I gained a reputation for meeting sales targets.

過去兩年我任職於一間知名貿易公司，並且是出了名的業務達人。

❻ I have ample work experience in related positions and consider myself well-equipped to fulfill the job's requirements.

我在相關職位擁有相當多的工作經歷，相信自己能夠達成此職位的需求。

結尾 Closing

❶ I can assure that I will contribute to the benefit of your company with my three-year experience in product marketing.

我相信以我在產品行銷三年的工作經驗，一定能對貴公司的收益帶來貢獻。

❷ I believe that my experience and professional skills would qualify me for this position.

我相信我的經驗以及專業技能使我具備勝任此職務的資格。

❸ The attached resume can provide you with more details of my professional background.

附件的個人履歷能提供您有關我專業背景的更多細節。

❹ Thank you for your time and consideration. I hope to have the opportunity to talk about the opening with you in person.

感謝您撥冗考慮。希望能有機會私下與您討論此職務。

❺ I am looking forward to a positive reply from your end.

我很期待能得到貴公司的正面回覆。

❻ Please see the attached resume for more details of my work experience.

關於我工作經歷的更多詳情，請參考附件的履歷。

✓ 不可不知的實用E-mail字彙

- **Fund Manager** 財務經理
- **Accounting Manager** 會計部經理
- **Accounting Staff** 會計部職員
- **Accounting Supervisor** 會計主管
- **Accountant** 會計
- **Accounting Assistant** 會計助理
- **Senior Accountant** 高級會計
- **Accounting Clerk** 記帳員
- **Auditor/Auditorial** 稽核人員
- **Personnel Manager** 人事部經理
- **Personnel Clerk** 人事部職員
- **Human Resources (HR)** 人力資源
- **Legal Adviser** 法律顧問
- **Procurement** 採購
- **Buyer** 採購員
- **Bond Analyst** 證券分析員
- **Bond Trader** 證券交易員
- **Trust Banking Executive** 銀行高級職員
- **Insurance Actuary** 保險公司理賠員

03 詢問面試結果

A Job Interview Follow-up Letter

 實際E-mail範例

     寄件者： Ginny

Dear Sir/Madam,

Introduction

In reference to my interview for the position of Sales Specialist dated Feb 25th, I am writing to enquire about your hiring decision.

Body

In view of the opportunity of the interview, I would like to express my interest in working with your company again.

Closing

Thank you for your time and consideration.

I would appreciate any feedback, no matter what your decision is.

Look forward to hearing back from you soon.

Best regards,
Signature of Sender
Sender's Name Printed

E-mail中譯　　寫信▼　　刪除　　回覆▼　　寄件者： Ginny

您好：

....

開頭

我於 2 月 25 日在貴公司面試業務專員一職，今日特地撰此信詢問您的聘雇決定。

本體

有鑒於此次面試機會，我希望能再次向您表達我想與您共事的意願。

結尾

感謝您的撥冗深思熟慮。

對於任何面試結果，我都能欣然接受。

期望能盡快得到您的回覆。

....

敬祝安康，
簽名檔／署名

 各段落超實用句型

說明：畫底線部分的單字可按照個人情況自行替換

開頭 Introduction

❶ **Regarding my interview for the position of <u>Project Manager</u> in your <u>Marketing Department</u> on <u>January 15th</u>, ...**

關於我在 <u>1 月 15 日</u>應徵貴公司<u>行銷部門專案經理</u>一職的面試，……

❷ **This letter serves to enquire about the outcome of my interview <u>last Monday</u>.**

這封信的目的是要詢問<u>上星期一</u>的面試結果。

❸ **After speaking with you and learning more about your company, I'm even more interested in working with you.**

在與您談話，並且對貴公司有更多認識後，我對於在貴公司工作更加感興趣了。

❹ At the end of the interview, you indicated that I should receive the outcome of the interview within two weeks.

在面試最後，您曾表示我會在兩星期內就收到面試結果。

❺ I would like to reiterate my interest in this position by means of this letter.

我希望能透過這封信件重申我對此職務的興趣。

❻ I am writing to inquire about the outcome of the interview we had two weeks ago, for the position of Hardware Engineer.

我寫這封信是希望瞭解我兩週前與您面試硬體工程師職位的結果。

本體 Body

❶ I am writing to let you know that I am very interested in the position of Senior Accountant that I have been interviewed for on March 10th.

我寫這封信的目的是要讓您知道我對 3 月 10 日所面試的資深會計師一職感到非常有興趣。

❷ I would like to enquire about the status of my application.

我想詢問目前有關我的工作應徵情形。

❸ I am very interested in working at CSBC Company and I am sure my experiences and skills would fit me for the position perfectly.

我很有興趣在 CSBC 公司工作。而且我相信以我的經驗及技能，我必定能理想地勝任此職務。

❹ I was very impressed with your company and its employment opportunities.

我對貴公司以及你們所提供的工作機會印象深刻。

❺ I would be very excited to have the opportunity to work with you.

若能擁有與您共事的機會我將會非常開心。

❻ I learned a lot about your company during the interview, and was highly impressed about your professionalism and friendly work environment.

在面試過程中，我更瞭解您的公司了，對您公司的專業度與友善的工作環境印象深刻。

結尾 Closing

❶ I hope the hiring process is going well.
希望聘雇過程進行順利。

❷ Please let me know if you need me to provide any information about myself to help you make the decision.
如果您需要我提供任何我的相關資訊，以協助您做出決定，請讓我知道。

❸ If necessary, I would be happy to provide any further information you need regarding my qualification and candidacy.
如果必要的話，我很樂意提供您所需任何有關我的候選資格及條件的進一步資訊。

❹ Please do not hesitate to contact me if you need any additional information.
如果您需要任何其他資訊，請不要客氣，儘管與我聯繫。

❺ I can be reached at 0912345678 or myname@gmail.com.
您可以透過電話 0912345678 或電子郵件 myname@gmail.com 與我聯繫。

❻ If you would like any other information for your decision-making process, please do not hesitate to let me know. I will be glad to supply anything you may need.
如果您在做出決定的過程中需要其他資訊，請不吝告知我。我很樂意提供任何您所需的資訊。

 # 不可不知的實用E-mail字彙

製造業
- **Product Manager** 生產部經理
- **Line Supervisor** 生產線主管
- **Research & Development Engineer** 研發工程師
- **Mechanical Engineer** 機械工程師
- **Production Engineer** 產品工程師

- **Maintenance Engineer** 維修工程師
- **Manufacturing Engineer** 製造工程師
- **Systems Operator** 系統操作員
- **Engineering Technician** 工程技術員
- **Programmer** 電腦程序設計師
- **Purchasing Agent** 採購員
- **Quality Control (QC) Engineer** 品質管理工程師

資訊人才
- **Computer System Manager** 計算機系統部經理
- **Management Information System (MIS)** 資訊管理系統
- **Computer Engineer** 計算機工程師
- **Hardware Engineer** 計算機硬件工程師
- **Product Support** 產品支援
- **Computer Data Input Operator** 計算機資料輸入員
- **Computer Processing Operator** 計算機處理操作員
- **Maintenance Engineer** 維修工程師
- **Technical Support Engineer** 技術支援工程師

業務人才
- **Sales Assistant** 業務助理
- **Sales Engineer** 業務工程師
- **Sales Executive** 業務主管
- **Sales Manager** 業務部經理
- **Sales Specialist** 業務專員
- **Real Estate Staff** 房地產專員

04 面試後感謝信

A Follow-up Thank You Letter

寫作要點 Key Points:

Step 1: 感謝對方給予面試機會。
Step 2: 表達希望得到職務的想法。
Step 3: 表示期待能儘快得到面試結果。

實際E-mail範例

| 寫信 ▼ | 刪除 | 回覆 ▼ | 寄件者: Ginny |

Dear Sir/Madam,

Introduction

Thank you for giving me the privilege of having an interview with you yesterday. I am grateful that you took much time out of your busy schedule to acquaint me with the company.

Body

After speaking with you, I am even more enthusiastic about working as a member of your team. With my qualification, skills and work experience, I consider myself suitable for this position.

Closing

I would like to thank you for your time and consideration again, and I do look forward to hearing for you regarding your hiring decision soon.

Best regards,
Signature of Sender
Sender's Name Printed

您好：

開頭

感謝您昨日給我可以與您面談的機會。

對於您百忙中撥冗讓我認識貴公司，我內心充滿感激。

本體

在與您談話過後，我對於能加入您的工作團隊更感熱切了。憑著我的條件資格、技能和工作經驗，我認為自己非常適任此職務。

結尾

我想再次感謝您的時間與考慮，而且我真的非常期待能很快就聽到關於您聘雇決定的消息。

敬祝安康，
簽名檔／署名

各段落超實用句型

說明：畫底線部分的單字可按照個人情況自行替換

開頭 Introduction

❶ Thank you for taking the time to meet with me last Friday.

感謝您撥空於上星期五與我見面。

❷ Thank you for taking time this morning to interview me for the position of Maintenance Engineer.

感謝您今日上午撥冗與我進行維修工程師一職之面試。

❸ I appreciated the opportunity to meet with you and your staff.

我很感謝能有機會與您以及您的職員見面。

❹ I am grateful to have the opportunity to meet with you and discuss the position of HR Manager this past Wednesday.

我很感謝能在這星期三有機會與您面談，並就人力資源經理一職做討論。

❺ I am writing to express my sincere gratitude for the opportunity of my interview <u>this Wednesday morning</u>.

我寫這封信是為了表達我對<u>本週三上午</u>的面試機會的誠摯感謝。

本體 Body

❶ It was very nice to discuss the expectations your company holds for the <u>Marketing Manager</u> position and my suitability for the position.

很高興能與您談論貴公司對<u>行銷經理</u>這個職務的期待，以及我對這個職位的適任度。

❷ The job description you made convinced me that this is a job I would enjoy.

您對這份工作內容所作的描述，使我相信這將會是個讓我勝任愉快的工作。

❸ I was very impressed with the professionalism of your organization and your staff.

我對貴公司以及您員工的專業印象深刻。

❹ The interview confirmed my interest in working with you.

這次面試確定了我想與您共事的意願。

❺ Again, I would like to convey my strong interest in joining your team.

我希望能再次傳達我想要加入您的團隊的強烈意願。

結尾 Closing

❶ I look forward to receiving the decision regarding my candidacy.

我期待能得知有關我的候選資格之決定。

❷ You suggested that the decision on the position will be made within <u>two weeks</u>, and I look forward to hearing your choice.

您告知有關該職位會在<u>兩週</u>內作出決定，而我很期待得知您的選擇。

❸ Please keep me informed as soon as your decision is made.

一旦您作出決定，請立刻通知我。

❹ I would be honored to start working with you and your team.

我將很榮幸能開始與您及您的團隊共事。

❺ I would like to reiterate my interest in the position and in working with you and your team.

我要再次表達我對此職務，以及與您及您的團隊共事的興趣。

 不可不知的實用E-mail字彙

- **Operational Manager** 業務經理
- **Business Controller** 業務主任
- **Business Manager** 業務經理
- **Customer Service Specialist** 客服專員
- **Product Marketing** 產品行銷
- **Public Relations** 公關
- **Advertising Staff** 廣告工作人員
- **Copywriter** 廣告文字撰稿人
- **Export Sales Manager** 外銷部經理
- **Export Sales Staff** 外銷部專員
- **Market Analyst** 市場分析員
- **Market Development Manager** 市場開發部經理
- **Marketing Manager** 行銷部經理
- **Marketing Staff** 行銷專員
- **Marketing Assistant** 行銷助理
- **Marketing Executive** 行銷主管
- **Marketing Representative** 行銷代表
- **Project Staff** 專案人員
- **Promotional Manager** 推廣銷售部經理

05 邀請面試告知信

An Interview Invitation Letter

寫作要點 Key Points:

Step 1: 感謝對方申請應徵此職務。
Step 2: 邀請對方參加面試，並告知時間、地點及其他注意事項。
Step 3: 提供聯絡方式，以便有問題時可以詢問。

實際E-mail範例

寫信 ▼	刪除	回覆 ▼	寄件者： Ginny

Dear Sir/Madam,

Introduction

Thank you for your interest in the position of Office Administrator in JSB Company.

Body

We would like to invite you to an interview for the position. Your interview has been scheduled for October 12th, 2017, 9 am, at No. 22, Sec. 2, Mingchuan E. Rd., Taipei.

Closing

If you have any questions or wish to reschedule, please call me at 02-1234-5678 or E-mail me at E-mailme@jsbcompany.com.

Best regards,
Signature of Sender
Sender's Name Printed

E-mail中譯

| 寫信 ▼ | 刪除 | 回覆 ▼ | 寄件者：Ginny |

您好：

開頭

謝謝您對 JSB 公司辦公室行政人員一職的興趣。

本體

我們想要邀請您來面試此職務。您的面試時間將排定在 2017 年 10 月 12 日上午 9 點。面試地點在台北市民權東路二段 22 號。

結尾

如果你有任何問題，或希望能更改面試時間，請致電 02-1234-5678，或來信 E-mailme@jsbcompany.com 與我聯絡。

敬祝安康，
簽名檔／署名

各段落超實用句型

說明：畫底線部分的單字可按照個人情況自行替換

開頭 Introduction

❶ Thank you for applying for the position of Export Sales Staff with our company.

感謝您應徵本公司的外銷部專員一職。

❷ We have received your application for the position of Sales Assistant. Thank you very much.

我們已經收到您對業務助理的應徵信函。非常感謝您。

❸ Thank you for submitting an online application for the position of Project Manager at HiBox Trading Company.

感謝您在線上遞交申請 HiBox 貿易公司專案經理一職的應徵信函。

❹ Following consideration of your application for the position of <u>HR Staff</u>, I am delighted to inform you that you have been short-listed for interview.

在看過您對<u>人力資源專員</u>一職的應徵信函後，我很高興通知您，您已被安排接受面試。

❺ We have received your application and CV. Thank you for applying for the position of <u>Public Relations</u>.

我們已經收到您的履歷與應徵信函，非常感謝您應徵本公司<u>公關</u>一職。

本體 Body

❶ We have reviewed your application and would like to invite you to have an interview with us.

我們已經看過您的應徵信，並希望能邀請您來與我們面試。

❷ Please visit our headquarters at <u>2 pm</u> on <u>September 19th, 2017</u> for an interview.

請於 <u>2017 年 9 月 19 日下午 2 點</u>到本公司<u>總部</u>進行面試。

❸ Your interview will be held in our head office, located at <u>7F, 212, Sec 3, Zhongxiao E. Rd., Taipei</u>.

您的面試將在我們位於<u>台北市忠孝東路三段 212 號 7 樓</u>的總公司舉行。

❹ Please visit our website, <u>www.hiboxtrading.com</u>, and schedule your interview.

請至本公司網站 <u>www.hiboxtrading.com</u> 預約您的面試時間。

❺ Please bring along the following documentation with you to the interview: <u>proof of ID</u>, <u>highest degree diploma</u> and <u>English proficiency certificate</u>.

請攜帶下列文件來參加面試：<u>身分證明文件</u>、<u>最高學歷畢業證書</u>及<u>英語能力證書</u>。

結尾 Closing

❶ If you have any difficulties scheduling your interview online, please contact me at <u>hello@hiboxtrading.com.tw</u> as soon as possible.

如果您無法在線上預定面試時間，請盡快來信 <u>hello@hiboxtrading.com.tw</u> 與我聯絡。

❷ **Please confirm your attendance by 12 pm on August 12th, 2017 by calling Jane 1234-5678 or E-mailing hello@hiboxtrading.com.tw.**

請在 2017 年 8 月 12 日中午 12 點之前，致電珍妮 1234-5678 或來信 hello@hiboxtrading.com.tw 以確認您會出席。

❸ **We look forward to meeting you.**

我們很期待能與您見面。

❹ **Please report to reception on arrival.**

抵達時請先至接待處報到。

❺ **If you are unable to attend the interview, please contact Jane by August 12th, 2017.**

如果您不克前來參加面試，請在 2017 年 8 月 12 日前與珍妮聯絡。

 **不可不知的實用E-mail字彙**

- **reception** 接待處
- **proof of ID** 身分證明文件
- **language proficiency certificate** 語言能力證明文件
- **diploma** 畢業證書
- **highest degree diploma** 最高學歷畢業證書
- **certificates of professional skills** 專業技能證照
- **2-inch certificate photo** 二吋證件照片
- **professional license** 專業證照
- **practical license** 執業證照
- **Police criminal record** 良民證
- **proof of legal residence** 合法居留證明

06 職位錄取告知信

A Job Offer Letter

寫作要點 Key Points:

Step 1: 通知對方已錄取職務。
Step 2: 告知對方報到時間，及相關注意事項。
Step 3: 提醒對方做出回覆，並表達歡迎之意。

實際E-mail範例

| 寫信｜▼ | 刪除 | 回覆｜▼ | 寄件者： Ginny |

Dear Sir/Madam,

Introduction

With regard to your application and the subsequent interview you had with us, we are pleased to offer you the position of Senior Accountant in JBS Company.

Body

Your first day in this position will be December 20th, 2017. Please find the attachment for the compensation package for this position.

Closing

Please let me know if you have any questions.

We are looking forward to your response.

Best regards,
Signature of Sender
Sender's Name Printed

E-mail中譯

寫信｜▼　刪除　回覆｜▼　寄件者： Ginny

您好：

開頭

有鑑於您的應徵申請書以及接下來與我們的面試，我們很高興通知您錄取 JBS 公司資深會計師一職的職務。

本體

您的報到日期為 2017 年 12 月 20 日。有關本職位的薪資方案請詳見附件。

結尾

如果有任何問題，請儘管問我。

我們很期待您的回音。

敬祝安康，
簽名檔／署名

各段落超實用句型

說明：畫底線部分的單字可按照個人情況自行替換

開頭 Introduction

❶ **On behalf of JBS Company, I am delighted to offer you the position of Marketing Planning Specialist.**

謹代表 JBS 公司，本人很高興能通知您錄取行銷策劃專員之職務。

❷ **I am pleased to inform you that we are extending you an offer of employment as Marketing Representative.**

很高興通知您，我們決定錄取您為行銷代表人員。

❸ **We are pleased to welcome you to join our team.**

我們很樂意歡迎您加入我們的團隊。

❹ Consequent to your prior interview and interaction with us, we are pleased to inform you that you have been selected for the Purchasing Manager position.

有鑑於您之前的面試以及與我們的互動交流，我們很高興通知您，您已被錄取為我們的採購經理。

❺ This letter serves to confirm that you're hired for the position of Purchasing Assistant with our company.

謹以此信確認您已受聘本公司採購助理之職務。

❻ We are pleased to inform you that our Board of Directors have decided to recruit you as part of our team.

我們很高興通知您，我們的董事會已決定僱用您加入我們的團隊。

本體 Body

❶ This is a full-time position of 40 working hours a week.

此乃一週 40 個工作時數的全職職務。

❷ Your initial salary will be NT$35,000 per month.

您的起始薪資將會是每個月新台幣三萬五千元。

❸ Please be noted that as a full-time employee, you are entitled to the standard benefits package of the company.

請注意身為全職員工，您享有本公司標準福利方案。

❹ Please prepare to begin work on November 1st, 2017.

請於 2017 年 11 月 1 日報到上班。

❺ The compensation for this position is attached.

本職位之薪資方案已隨信附件。

❻ You will be able to find the details concerning compensation and employee benefits in the attached file.

您可在附件中看到薪資與員工福利的相關細節。

結尾 Closing

❶ If you decide to accept this offer, please reply to this letter as a confirmation.

如果您決定接受此職務，請回覆此信作為確認。

❷ **Should you have any questions regarding employment policies and procedures, please do not hesitate to contact me.**

萬一您有任何關於雇用方針或程序上的疑問，請儘管與我聯繫。

❸ **Please confirm your acceptance of this offer by replying to this letter no later than October 20th, 2017.**

請在 2017 年 10 月 20 日之前回覆此信，以確認您接受此一職務。

❹ **Please get back to us with your decision on whether you would accept this job offer as soon as possible.**

無論您是否願意接受這份職務，煩請盡快回覆我們您的決定。

❺ **We are looking forward to your joining our staff.**

我們很期待您成為我們的一員。

❻ **We are excited to have you on board and will appreciate your reply at your earliest convenience.**

我們非常期待您加入我們，並希望能及早獲得您的回覆。

 **不可不知的實用E-mail字彙**

- **start date** 報到日期（開始上班日）
- **employment policies** 雇用方針
- **compensation package** 薪資方案
- **benefit package** 福利方案
- **full-time employment** 全職工作
- **initial salary** 起始薪資
- **annual salary** 年薪
- **health insurance** 健康保險
- **labor insurance** 勞工保險
- **three Chinese festivals bonus** 三節獎金
- **year-end bonus** 年終獎金
- **annual leave** 年假

Note　寫信｜▼　刪除　回覆｜▼　寄件者：

Part2

各式介紹篇

01 向同事自我介紹

A Self-Introduction Letter as A New Employee

寫作要點 Key Points:

Step 1: 介紹自己的姓名，並說明來信目的。
Step 2: 說明受僱的工作職稱及負責的工作項目，並提供聯絡方式。
Step 3: 表示期待與對方共事，並展現友善態度。

實際E-mail範例

寫信｜▼	刪除	回覆｜▼	寄件者：	Jennie

Dear Mr./Ms./Dr./Professor Surname, （對上級）

Dear Colleagues,/ Hello, Name, （對平輩同事）

Introduction

My name is Jennie Chen. I am writing this letter to introduce myself as your new colleague in the Accounting Department of JSB Company.

Body

I have been employed as an Accounting Assistant and will be working together with you for the length of my contract.

Closing

If I can be of service at any moment, please feel free to contact me at my extension 3023.

Best regards,
Signature of Sender
Sender's Name Printed

E-mail中譯　　寫信｜▼　刪除　回覆｜▼　寄件者：Jennie

親愛的～先生／女士／醫生／博士您好：（對上級）
親愛的同事們 / 姓名您好：（對平輩同事）

開頭

我是陳珍妮。謹以此信向各位自我介紹，我是 JSB 公司會計部門的新進同仁。

本體

我目前擔任會計助理一職，並會在合約期間與各位共事。

結尾

如果任何時候需要我的服務，請撥打我的分機號碼 3023 與我聯絡。

敬祝安康，
簽名檔／署名

各段落超實用句型

說明：畫底線部分的單字可按照個人情況自行替換

開頭 Introduction

❶ My name is Emily Chung. I have just been hired as the new Sales Manager here at JSB.

我的姓名為鍾愛蜜麗。甫受僱為 JSB 公司的新任業務經理。

❷ I wish to take a moment to introduce myself to you by this email.

我希望能花一分鐘的時間，透過這封郵件向各位自我介紹。

❸ I am sending this email to you with the purpose of introducing myself to you.

我寄這封信給各位的目的，是為了向大家做自我介紹。

❹ I would like to take this opportunity to introduce myself as the new Marketing Analyst of JSB Company.

我想借此機會，以 JSB 公司的新任市場分析師的身份，向各位自我介紹。

❺ I have joined **JSB Company** recently. I am honored to be able to be part of the team.

我近日已經加入 JSB 公司。很榮幸能成為團隊的一員。

本體 Body ┈┈┈┈┈┈┈┈┈┈┈┈┈┈┈┈┈┈┈┈┈┈┈┈┈┈┈┈┈┈┈┈┈┈

❶ I have been assigned to the **Marketing Department** and will be in charge of handling the **sales and marketing planning**.

我被分派到行銷部門，並將負責業務及行銷策劃。

❷ I will be taking the role of **Director of the RD Department**, starting on **May 10th**.

我將從 5 月 10 日起，擔任研發部門的經理一職。

❸ I am hired to assist you with all kinds of office work.

我受聘協助各位各項辦公室工作。

❹ I have been appointed as **the director** of the **Research & Development Department**.

我接受任命成為研發部門的主管。

❺ I am excited about joining the team and working together with you.

我對於能加入這個團隊，並且能與各位共事，感到非常興奮。

結尾 Closing ┈┈┈┈┈┈┈┈┈┈┈┈┈┈┈┈┈┈┈┈┈┈┈┈┈┈┈┈┈┈┈┈

❶ If you have any questions or would like to meet me in person, please feel free to stop by my office.

如果您有任何問題，或想要親自與我碰面，歡迎您到我辦公室來。

❷ Should anyone need to reach me for any reason, my extension number is **3023**.

若是有任何原因需要找我，我的分機號碼是 3023。

❸ If anyone wants to reach me, just call me at my extension **3023** or message me at **myname@jsb.com**.

如果任何人需要找我，只要打我分機 3023 或是寫信到 myname@jsb.com 給我就可以了。

❹ I look forward to meeting all of you in person at the Newcomers' Welcome Party on **Friday night**.

我很期待能在星期五晚上的迎新派對上與各位見面。

❺ Thank you for welcoming me to this company.
謝謝你們歡迎我加入這個公司。

 不可不知的實用 E-mail 字彙

- **Head Office** 總公司
- **Branch Office** 分公司
- **Chairman's Office** 董事長辦公室
- **General Manager's Office** 總經理辦公室
- **Administration Department** 管理部／秘書室
- **Human Resources Department (HR)** 人力資源部
- **General Affairs Department** 總務部
- **General Accounting Department/Finance Department** 財務部
- **Sales Department** 業務部
- **Planning Department** 企劃部
- **Product Development Department** 產品開發部
- **Research and Development Department (R&D)** 研發部
- **Purchasing/Procurement Department** 採購部
- **Engineering Department** 工程部
- **Marketing Department** 行銷部
- **Customer Service Department** 客服部

02 向廠商自我介紹

A Self-introduction Letter to Clients

寫作要點 Key Points:

Step 1: 說明來信目的：自我介紹。
Step 2: 介紹自己所負責的工作項目，並對合作採取主動態度。
Step 3: 表示希望建立良好合作關係，並感謝對方支持。

實際E-mail範例

寫信 ▼	刪除	回覆 ▼	寄件者： Ginny

Dear Sir/Madam,

Introduction

Please let me take this opportunity to introduce myself as the newly appointed Sales Representative for JSB Trading Ltd.

Body

I would like to set up a personal meeting with you at your convenience, to discuss the current needs and concerns of your company and at the same time review JSB's latest offering of products and services.

Closing

I will contact your assistant tomorrow to make an appointment with you. My hope is to meet at your office by the end of this month.

Best regards,
Signature of Sender
Sender's Name Printed

E-mail中譯　[寫信|▼]　[刪除]　[回覆|▼]　寄件者：[Ginny]

您好：

開頭

請容許我藉這個機會向您自我介紹，我是 JSB 貿易股份有限公司新任的業務代表。

本體

我希望能在您方便的時間與您安排一次個人會面，以討論貴公司目前的需要及考量，並且同時向您介紹 JSB 最新的產品及服務項目。

結尾

我將會在明天與您的助理聯繫，以預約和您碰面的時間。我希望能夠在本月底之前到您的公司與您見面。

敬祝安康，
簽名檔／署名

各段落超實用句型

説明：畫底線部分的單字可按照個人情況自行替換

開頭 Introduction

❶ **The purpose of this letter is to briefly introduce myself to you as JSB's new Sales Representative.**

這封信的目的是要向您簡短自我介紹，我是 JSB 公司的新任業務代表人員。

❷ **Please allow me to introduce myself. I am Alex Cooper, the Sales Representativeof JSB Company.**

請容許我向您自我介紹。我是艾利克斯庫柏，JSB 公司的業務代表。

❸ **My name is Tracy Cheng, and I am the Chief Editor at Fun Children's Books.**

我是程崔西，是歡樂童書的總編。

❹ I will be your Project Manager for the software development project.

我將會是負責您軟體開發專案的專案經理。

❺ My name is Larry Fang, the chief planner of your company's year-end party.

我是方賴瑞，負責貴公司年終晚會活動的主要策劃人。

本體 Body

❶ I would like to pay you a visit at your convenience.

我希望能在您方便的時間前去拜訪您。

❷ I would like to schedule a meeting with you at your office to discuss the details of the development project.

我希望能安排時間到您公司與您會面，以討論這次專案開發的細節。

❸ I will contact your secretary by the end of this week to make an appointment with you.

我會在這週結束之前聯絡您的秘書，預約與您見面的時間。

❹ It will be great if we could have a discussion about your project by the end of this month.

如果在本月底之前，我們可以針對您的專案做討論，就太棒了。

❺ I would like to meet you in person and show you the samples of our latest products.

我希望能親自拜訪您，並讓您看看本公司最新產品的樣品。

結尾 Closing

❶ I am looking forward to seeing you soon.

我期待能很快與您見面。

❷ Please let me know when the best time to visit you at your office is.

請讓我知道何時是到公司去拜訪您的最佳時間。

❸ I will get in touch with your personal assistant by Friday, and set up a time to meet with you.

我將在本週五前與您的私人助理聯絡，並決定與您會面的時間。

❹ We can have a luncheon and discuss your needs and concerns about this project.

我們可以安排一次工作午餐，討論您對這次專案的需求與關切重點。

❺ Please do not hesitate to contact me if you need any additional information.

如果您需要任何其他的資訊，請儘管與我聯繫。

 # 不可不知的實用**E-mail**字彙

- **arrange a meeting** 安排碰面
- **schedule a meeting** 排定會面時間
- **make an appointment** 預約時間
- **set up a time to meet** 決定見面時間
- **personal meeting** 個人會議
- **luncheon** 工作午餐，午餐會
- **pay a visit to sb.** 拜訪某人

- **cancel a meeting** 取消會面
- **reschedule a meeting** 重新安排會面時間

03 向廠商介紹公司

A Company Introduction Letter

寫作要點 Key Points:

Step 1: 說明來信目的為介紹公司。
Step 2: 簡短陳述公司的主要營運項目。
Step 3: 主動提出期盼未來能夠合作。

 實際E-mail範例

  　　　　寄件者：　Ginny

Dear Mr./Ms./Dr./Professor Surname,

Introduction

I would like to take an opportunity to introduce our Company to you. JSB Trading Ltd. has been engaged in import and export business for the last 15 years.

Body

We specialize in the import and export of household appliances among Europe, America and Asia, and we have gained a fine reputation for our excellent services.

Closing

We would like to offer our services to your department store. Our Marketing Manager will contact you shortly and provide you a detailed description of our services.

We are looking forward to working with you.

Best regards,
Signature of Sender
Sender's Name Printed

E-mail中譯

寫信｜▼　　刪除　　回覆｜▼　　寄件者：Ginny

親愛的〜先生／女士／醫生／博士您好：

開頭

我希望能用這個機會向您介紹本公司。JSB 貿易有限公司在過去 15 年來從事進出口業務。

本體

我們專門從事歐洲、美洲及亞洲之間的家電進出口，並且因為服務卓著而在業界享有極佳的聲譽。

結尾

我們很希望能為貴百貨公司提供服務。我們的行銷經理將會很快與您聯繫，以提供您有關本公司服務項目的更詳細説明。

我們很期待能與您合作。

敬祝安康，
簽名檔／署名

各段落超實用句型

說明：畫底線部分的單字可按照個人情況自行替換

開頭 Introduction

❶ **Please allow me to take this opportunity to introduce our company.**
請容許我借此機會向您介紹本公司。

❷ **I am writing this letter to introduce our company and our services to you.**
謹以此信向您介紹本公司以及本公司的服務項目。

❸ **I am writing to you on behalf of JSB Trading Ltd.**
本人謹代表 JSB 貿易股份有限公司寫這封信給您。

❹ Our company has been involved in the <u>event organizing business</u> for the past <u>six years</u>.

過去六年來，本公司一直從事活動策劃業務。

❺ <u>JSB</u> has been engaged in the <u>export of household appliances</u> since it was established in <u>2003</u>.

<u>JSB</u> 公司自 <u>2003</u> 年成立以來，一直從事家電器用品外銷的工作。

本體 Body

❶ I would like to pay you a visit in order to present the services we provide in detail.

我希望能拜訪您，向您詳細解釋我們所提供的服務內容。

❷ I would like to meet with you in person and offer more detailed information about our products.

我希望能私下與您碰面，以提供有關本公司產品更詳細的資訊。

❸ Our product catalogue is attached for your consideration.

謹隨信附上本公司產品目錄，供您參考。

❹ Our <u>Sales Specialist</u>, <u>Mr. Wu</u>, will get in touch with you by <u>the end of this week</u>.

我們的業務專員吳先生會在這週結束之前與您聯繫。

❺ We provide high quality services for reasonable prices.

我們以合理的價格，提供高品質的服務。

結尾 Closing

❶ Please feel free to contact me if you have any questions regarding our services.

若您對本公司的服務項目有任何問題，歡迎與我聯絡。

❷ Please let me know if you need any further information regarding our products.

若您需要有關本公司產品的進一步資訊，請與我聯繫。

❸ Product samples will be sent to you on demand.

索取後產品樣本將會寄送給您。

❹ **We would like to provide the services of our company to your underline{restaurant}.**
我們希望能為貴餐廳提供本公司的服務。

❺ **We look forward to working with you in the near future.**
我們期待能在不久的將來與您合作。

 # 不可不知的實用E-mail字彙

- **event planning** 活動企劃
- **wedding arrangement** 婚禮策劃籌備
- **manufacture of computer components** 電腦零件製造
- **manufacture of household appliances** 家電用品製造
- **manufacture of electronic components** 電子零件製造
- **export of household appliances** 家電外銷
- **import of agricultural products** 農產品進口
- **web design** 網頁設計
- **website architecture** 網頁架構
- **clothing export** 成衣外銷
- **real estate investment** 房地產投資
- **real estate development** 房地產開發
- **interior design** 室內設計
- **product packaging design** 產品包裝設計

04 向廠商介紹產品／服務

A Product Introduction Letter

實際E-mail範例

寫信 ▼ ┊ 刪除 ┊ 回覆 ▼ ┊ 寄件者： Ginny

Dear Sir/Madam,

Introduction

On behalf of JSB Company, I am happy to take this opportunity to introduce you the latest services of JSB Company.

Body

Our company has been engaged in event arrangements for the past 15 years and has been providing our services to large business organizations at competent prices. Our services include planning large events such as opening ceremonies and year-end parties as well as private parties such as wedding parties and birthday celebrations.

Now we have expanded our business and currently provide activity consultation services and product promotion services.

Closing

I attached the catalogue of our services to this mail for your consideration.

Should you have any questions regarding our services, please do not hesitate to contact me by calling 1234-5678 or emailing myname@gmail.com.

We look forward to providing you with our high quality services in the near future.

Best regards,
Signature of Sender
Sender's Name Printed

E-mail中譯

寫信｜▼ 刪除 回覆｜▼ 寄件者：Ginny

您好：

開頭

本人謹代表 JSB 公司，很高興利用這個機會向您介紹我們最新的服務項目。

本體

本公司在過去十五年來從事活動策劃統籌，並且以極具競爭性的價格為許多大型企業提供我們的服務。我們的服務項目包括大型活動，如開幕典禮和年終晚會，以及私人派對如結婚派對和慶生派對等。

目前我們已經拓展我們的業務範圍，並開始提供活動諮詢服務，以及產品促銷活動服務。

結尾

隨信附上我們的服務項目型錄，以供您參考。

萬一您有任何有關我們服務項目的問題，請儘管撥打1234-5678或來信myname@gmail.com與我聯絡。

希望很快能有機會為您提供本公司高品質的服務。

敬祝安康，
簽名檔／署名

各段落超實用句型

說明：畫底線部分的單字可按照個人情況自行替換

開頭 Introduction

❶ I am writing this letter on behalf of <u>JSB Ltd.</u> to introduce you the latest products of our company.

謹代表 <u>JSB 股份有限公司</u>向您介紹本公司的最新產品。

❷ By means of this letter, I would like to introduce our new products to you.

我希望能藉由此信向您介紹本公司的新產品。

❸ It is with great pleasure to announce that we have launched a new line of cosmetics.

很高興向您宣布，本公司已經推出新系列的化妝品。

❹ We take our opportunity with this letter to introduce you one of our latest products.

我們要利用這封信的機會向您介紹我們最新產品之一。

❺ We are proud to inform you that our latest products have recently been launched.

我們很榮耀地在此通知您，本公司的最新產品已於近日推出上市。

本體 Body

❶ We have experience in organizing different types of events from formal business conferences to private parties.

我們有統籌企劃各種類型活動的經驗，包含正式的商務會議到個人派對。

❷ We specialize in wedding arrangements, from wedding ceremonies to wedding banquets.

我們專門從事婚禮籌備工作，籌備項目包含結婚儀式到結婚喜宴。

❸ I have attached the latest catalogue of our products and services to this letter for your consideration.

我已經隨信附上本公司最新產品及服務型錄，以供您參考。

❹ Please find the attachment for our product catalogue.

請打開附件，參看本公司的產品型錄。

❺ I will be happy to meet you in person and provide detailed information about our products and services.

我很樂意親自拜訪您，提供您本公司產品及服務項目的詳細資訊。

結尾 Closing

❶ It is our sincere hope to establish a partnership with you.

我們誠摯希望能與貴公司建立合作夥伴關係。

❷ We are happy to provide you with samples to try out.

我們很樂意提供您樣品試用。

❸ I would call up your office to fix an appointment and I look forward to meeting you in person.

我會致電您的辦公室安排並會面時間，期待能親自與您碰面。

❹ Please let me know if you need any further information from our end.

如果您需要我方提供任何進一步資訊，請與我聯絡。

❺ We are looking forward to working with you in the future.

我們期待將來能與您合作。

不可不知的實用E-mail字彙

- **product catalogue** 產品目錄
- **list of services** 服務項目清單
- **sample product** 產品樣本
- **additional service** 額外服務
- **trial product / sample** 試用品

A Letter Introduce a New Colleague to Clients

寫作要點 Key Points:

Step 1: 說明來信目的為介紹新同事或團隊。
Step 2: 說明新同事所負責的工作項目，並給予支持。
Step 3: 感謝對方，並表達能維持合作的希望。

 實際E-mail範例

寫信｜▼	刪除	回覆｜▼	寄件者： Ginny

Dear Sir/Madam,

Introduction

I am very happy to take this opportunity to introduce our new Marketing Analyst Nelson Tam. We are excited to have Nelson join our team. I believe you will enjoy reaping the benefits of his professional knowledge and experience in market analysis.

Body

Let us know whenever you need a brand strategy or marketing consulting, and we will arrange a time for you to meet with Nelson.

Closing

Thank you for your business and we look forward to seeing you soon.

Best regards,
Signature of Sender
Sender's Name Printed

E-mail中譯

[寫信｜▼] [刪除] [回覆｜▼] 寄件者：[Ginny]

您好：

開頭

很高興能藉此機會向您介紹我們的新市場分析師——譚尼爾森。我們對於尼爾森能加入我們的團隊感到非常興奮。我相信您將會受益於他在市場分析方面的專業知識與經驗。

本體

無論何時，當您需要品牌策略或市場諮詢，請與我們連繫，我們會安排您與尼爾森會面的時間。

結尾

感謝您的合作，希望能很快與您見面。

敬祝安康，
簽名檔／署名

各段落超實用句型

說明：畫底線部分的單字可按照個人情況自行替換

開頭 Introduction

❶ It is with great pleasure that I introduce to you our new Sales Specialist, Jennifer Miller.
非常高興能向您介紹我們的新業務專員——珍妮佛米勒。

❷ By means of this letter, I would like to introduce Sam Wang to you.
藉由此信，我想向您介紹王山姆。

❸ We are excited to have Peter join our team.
我們對於彼得能加入我們的團隊感到非常興奮。

❹ We are delighted to inform you that John Jones has been selected to be our new Sales Representative.
很高興通知您，約翰瓊斯已經獲選成為我們的新業務代表。

❺ Please allow me to introduce our new colleague, Sam Wang, who has officially joined us as an Accounting Assistant on November 1st, 2017.

請容許我向您介紹我們的新同仁王山姆，他已經在 2017 年 11 月 1 日以會計助理一職，正式加入我們。

❻ I am writing to introduce the newest addition to our team, May Huang, who has extensive experience in Sales.

我寫這封信是要介紹本團隊的新成員黃梅，她在業務方面擁有相當豐富的經驗。

本體 Body

❶ Lydia Ling will replace Judy Lee to be in charge of the export business of our company from November 1st, following Ms. Lee's resignation.

林莉雅將取代李茱蒂，在李小姐離職後於 11 月 1 日開始負責本公司的出口業務。

❷ Mr. Wang will be handling most of our accounts including yours from next month.

王先生將從下個月開始負責本公司大部分帳務，包含貴公司的帳務。

❸ Ms. Lai has replaced Mr. Peng, who is no longer in our employment.

賴小姐已經取代目前已離職的彭先生。

❹ Mr. Guo has extensive experience in event planning for over ten years.

郭先生在活動企劃方面有十年以上的廣泛經驗。

❺ Ms. Smith will take over responsibility for the technical management and technical maintenance.

史密斯先生將承接技術管理及技術維修的職務。

❻ Mr. Chou will be in charge of all account management affairs and related procedures.

周先生將負責所有與帳戶管理相關的事務與程序。

結尾 Closing

❶ We are confident that Ms. Julie Song will be an asset to both our company and our clients.

我們深信宋茱莉小姐對本公司及我們的客戶，都會是一項資產。

❷ I believe that <u>Ms. Song</u> will be able to provide you with the same quality service as her predecessor.

我相信<u>宋小姐</u>將會如前任一般提供您相同的高品質服務。

❸ We look forward to your support and cooperation with <u>Mr. Jones</u> in his current assignment.

我們期待您能在<u>瓊斯先生</u>任職期間給予支持與合作。

❹ We thank you for your continuous support and cooperation.

感謝您繼續支持與合作。

❺ Should you have any questions regarding this change, please feel free to let us know.

若您對此異動有任何疑問,請儘管與我們聯繫。

 不可不知的實用E-mail字彙

- **take over the responsibility** 接任職務
- **replace sb.** 取代某人
- **predecessor** 前任,前輩
- **following sb.'s resignation** 在某人離職之後
- **succeed sb. in a post** 繼任某人的職務
- **continuous support** 繼續支持

Part3
簡易提案篇

01 向廠商提出方案

A Business Proposal Letter

寫作要點 Key Points:

Step 1: 向對方提出方案。
Step 2: 簡短說明提案內容。
Step 3: 表達希望能進一步討論提案。

實際E-mail範例

| 寫信▼ | 刪除 | 回覆▼ | 寄件者： | Ginny |

Dear Sir/Madam,

Introduction

In view of the conversation we had last week, I realize that your office expenses has been a high-priority problem for you. Hence I would like to propose a cost-effective solution to help you minimize the in-house burden.

Body

We are a company that specializes in accounting services. We provide our clients with high-quality professional services and, of course, the highest level of confidentiality.

Closing

Please contact me directly if you have additional questions or requests.

I look forward to discussing this proposal in detail with you soon.

Best regards,
Signature of Sender
Sender's Name Printed

E-mail中譯　　寫信 ▼　　刪除　　回覆 ▼　　寄件者：Ginny

您好：

....

開頭

鑑於我們上週的談話，我明白辦公室支出是貴公司目前最大的問題。因此我想提出一個成本效益解決方案來協助您將內部負擔降至最低。

本體

我們是一間專門從事會計帳目服務的公司。我們提供客戶高品質的專業服務，以及最高等級的保密。

結尾

如果您有其他問題或請求，請直接與我聯繫。

期待能很快能與您詳細討論此專案。

....

敬祝安康，

簽名檔／署名

各段落超實用句型

說明：畫底線部分的單字可按照個人情況自行替換

開頭 Introduction

❶ **I am writing to you with a business proposal to help you increase your sales.**

我寫這封信是要向您提出一個可以幫助您提高營業額的商業提案。

❷ **I know that you are looking for business expansion in China; therefore, I would like to propose a business expansion plan here.**

得知您目前正尋求在中國拓展業務，因此我想在此提出一個業務擴增的方案。

❸ **I am writing this letter to propose a product promotion plan for the launch of your new products.**

謹以此信向您提出一個針對貴公司新上市產品而設計的產品促銷方案。

❹ By means of this letter I am pleased to offer a proposal that is especially designed to deal with your <u>manpower shortage</u>.

很高興藉由此信向您提出一個特別為應付貴公司<u>人力短缺問題</u>而設計的提案。

❺ In view of our discussion on <u>May 2nd</u> with you, our team has designed a proposal to <u>deal with the excess stock in your inventory</u>.

有鑑於與貴公司在 5 月 2 日所做的討論，我們團隊已經為您設計出一個<u>處理倉庫存貨過多</u>的方案。

本體 Body

❶ Our company has been a pioneer in <u>providing innovative promotion ideas to boost sales of products</u>.

本公司在<u>提供創新促銷點子以提升產品銷售方面</u>，一直都是業界先驅。

❷ We have managed to cover every area of concern in this proposal that will benefit your existing business.

我們已經設法將每個有益於您目前業務的考量點涵蓋在這份提案中。

❸ This proposal is designed to increase your <u>productivity</u>.

這份提案是為了提升貴公司<u>產品產量</u>而設計的。

❹ In the attached outline proposal you will find all the relevant information you need.

在附件的提案大綱中你將會找到你所需要的所有相關資訊。

❺ In the proposal we have outlined how your company can <u>attract new clients and retain existing customers</u> at the same time.

在這份提案中，我們列出了貴公司可以<u>吸引新客戶並同時保留原客戶</u>的方法。

結尾 Closing

❶ I attached the outline proposal for your review.

我將提案大綱隨信附上，以供您詳閱。

❷ Please feel free to contact me if you have any questions regarding our business proposal.

如果您有任何關於這份商業提案的問題，請儘管與我聯繫。

❸ **Please contact me after reviewing the proposal and discuss with me whether it needs modification.**
請在看完提案後與我聯絡，討論內容是否有需要修改。

❹ **I would like to set up a time to meet with you to discuss the details of implementation of the proposal.**
我希望能跟您安排一個會面時間，以討論提案執行的細節。

❺ **I am looking forward to receiving your feedback.**
我很期待能收到您對這份提案的回應。

 # 不可不知的實用E-mail字彙

- **high-priority problem** 最優先的問題
- **office expense** 辦公室支出
- **in-house burden** 內部負擔
- **marketing cost** 行銷成本
- **delivery cost** 配送成本
- **agency cost** 業務成本
- **operating expense** 營業費用
- **repair and maintenance expense** 修繕費用
- **office supplies expense** 辦公室文具費用
- **traveling expense** 差旅費
- **Research and Development expense** 研發費用
- **loss on physical inventory** 存貨盤損
- **excess stock in inventory** 庫存過多
- **manpower shortage** 人力短缺

02 贊成廠商的方案

A Proposal Approval Letter

寫作要點 Key Points:

Step 1: 感謝對方提出方案。
Step 2: 表示對提案的認同與讚賞。
Step 3: 表達合作意願（討論後續執行或簽約）。

實際E-mail範例

| 寫信 ▼ | 刪除 | 回覆 ▼ | 寄件者： Ginny |

Dear Sir/Madam,

Introduction

Thank you for the business proposal which you sent to us on August 20th, 2017.

Body

We are very pleased to inform you that we have accepted your proposal for our business expansion plan. Your proposal covers all aspects of our requirements. Our Board of Directors is very satisfied with your understanding of this project.

Closing

To get this project underway immediately, we would like to discuss details with you and have the necessary paperwork signed.

Please contact me at your convenience to schedule a personal meeting.

Best regards,
Signature of Sender
Sender's Name Printed

E-mail中譯

寫信▾　刪除　回覆▾　寄件者：Ginny

您好：

開頭

感謝您於 2017 年 8 月 20 日寄給我們的商業提案。

本體

我們很高興通知您，我們已經接受了您對本公司業務擴充計劃的提案。您的提案涵蓋了我們所要求的每一項條件。本公司董事會非常滿意您對於本專案的了解程度。

結尾

為了能讓此專案立刻進行，我們希望能與您討論細節，並簽署必要文件。

請在您方便的時間與我聯絡，以安排碰面時間。

敬祝安康，
簽名檔／署名

 各段落超實用句型

說明：畫底線部分的單字可按照個人情況自行替換

開頭 Introduction

❶ **Thank you for contacting us for your business proposal.**

謝謝您聯繫本公司，並提供您的商業提案。

❷ **I would like to thank you for offering us the proposal to assist us in business expansion.**

我要感謝你們提供協助我們拓展業務的提案。

❸ **In reference to the business proposal that you have sent to us on August 20ᵗʰ, 2017, ...**

關於你們在 2017 年 8 月 20 日寄給我們的商業提案，……

④ We have reviewed the promotion proposal that you designed for our newly launched product.

我們已經看過貴公司為我們新上市產品所設計的促銷方案。

⑤ In response to your business proposal letter dated on July 18th, 2017, ...

回覆您在 2017 年 7 月 18 日的商業提案信件，……

本體 Body

❶ I am pleased to inform you that your proposal has been approved by our Administration Department.

很高興在此通知您，您的提案已經通過本公司管理部的審核了。

❷ It is with great pleasure to confirm our acceptance of your proposal.

非常高興在此跟您確定我們決定接受您的提案。

❸ We are happy to inform you of the approval of your business proposal.

我們很高興通知您，您的商業提案已獲認同。

❹ We are glad to inform you that we have accepted your proposal for our new office building construction project.

我們很高興通知您，我們已經接受了您對本公司新辦公大樓的建設方案。

❺ We were impressed with your comprehensive understanding of the scope of the project.

您對這項專案領域的了解讓我們印象深刻。

❻ We admire your confidence in being able to complete the project on schedule.

我們對您有自信能如期完工感到非常讚賞。

結尾 Closing

❶ We need to schedule a meeting and begin discussing details immediately.

我們需要安排一次會議，並立刻開始討論細節。

❷ We would like to arrange a meeting to review the project specifications.

我們想要安排一次會議，檢視專案的具體內容。

❸ **We have to set up a meeting and sign the necessary paperwork.**
我們必須安排一次會面，並簽署必要文件。

❹ **Please contact me at your convenience to fix a meeting.**
請在您方便的時間與我聯繫，好安排會面。

❺ **If you find this proposal favorable, please let me know as soon as possible and we can arrange for a time and location to meet for further discussions.**
若您覺得此提案不錯，請盡快告知我，我們可以安排時間與地點更進一步討論。

 # 不可不知的實用E-mail字彙

- **business expansion proposal** 業務擴充方案
- **product promotion proposal** 產品促銷方案
- **new product launch proposal** 新品上市方案
- **sales increase proposal** 業務提升方案
- **cost effective solution** 具成本效益的提案
- **office building expansion proposal** 辦公大樓擴充提案
- **factory expansion proposal** 廠房擴充提案

03 拒絕廠商的方案

A Proposal Declination Letter

 實際E-mail範例

| 寫信 ▼ | 刪除 | 回覆 ▼ | 寄件者： Ginny |

Dear Sir/Madam,

Introduction

In reference to the letter you sent to us on October 21st, 2017, we thank you for submitting your proposal.

Body

However, it is with great regret that I have to inform you that we have to decline your proposal, as business expansion is not a high priority of our restaurant for the time being.

Closing

We thank you again for your proposal and we do hope to have opportunity to work with you in the future.

Best regards,
Signature of Sender
Sender's Name Printed

E-mail中譯

寫信▼　刪除　回覆▼　寄件者：Ginny

您好：

開頭

我們很感謝您在 2017 年 10 月 21 日寄給我們的信件中的提案。

本體

然而，很遺憾地必須通知您，我們必須婉拒您的提案，因為擴展業務並不是本餐廳目前最優先考量的事。

結尾

再次感謝您的提案。我們真心希望未來能有與您合作的機會。

敬祝安康，
簽名檔／署名

各段落超實用句型

說明：畫底線部分的單字可按照個人情況自行替換

開頭 Introduction

❶ **We appreciate your business proposal letter dated September 19ᵗʰ, 2017.**

很感謝您 2017 年 9 月 19 日這天寄來的商業提案信。

❷ **In answer to your business proposal letter dated September 19ᵗʰ, 2017, ...**

謹回覆您於 2017 年 9 月 19 日寄來的商業提案信，……

❸ **Thank you for submitting your business proposal regarding our factory expansion plan.**

感謝您提交有關本公司廠房擴充計劃的商業提案。

❹ **We have received your promotion proposal for our new line of products.**

我們已經收到您對本公司新系列產品的促銷提案。

❺ Regarding the proposal you submitted last Friday, ...

關於您上週五提交的方案，⋯⋯

❻ This is regarding the proposal you have emailed us yesterday. My colleagues and I have gone through the proposal thoroughly and would like to inform you of our decision.

這封信是關於您昨天寄給我們的提案。我的同事與我已經徹底閱讀過提案，想要告知您我們的決定。

本體 Body

❶ I am writing to inform you that the proposal has been disapproved.

謹以此信通知您，您的提案未經通過。

❷ It is with regret that we need to decline this proposal.

我們必須很遺憾地婉拒這份提案。

❸ Unfortunately, we have to decline this proposal.

很遺憾地，我們必須婉拒這份提案。

❹ We are not ready for business expansion at the moment.

我們目前還沒準備好要擴充營業。

❺ The proposal you submitted does not meet our current needs.

您所提出的提案並不符合我們目前的需求。

❻ While we loved the proposal, we regret to say that our current budget does not allow us to accept it.

雖然我們非常喜歡這個提案，很遺憾的是，我們目前的預算不允許我們接受此提案。

結尾 Closing

❶ Thank you again for submitting your proposal.

再次感謝您提交提案。

❷ We are sorry we are not able to work with you this time.

很遺憾這次無法與你們合作。

❸ We thank you for your time and effort.

感謝您所付出的時間與努力。

❹ **We look forward to any possible opportunity to work with you in the future.**
我們期待未來任何可能與您合作的機會。

❺ **We will definitely get in touch with you if we have other business plans.**
如果我們有其他商業計劃，一定會與您聯絡。

❻ **We are very grateful for the thought and time you have put into the proposal, and truly appreciate your hard work.**
我們非常感激您在此提案上花費的心力與時間，也很感謝您的努力。

 # 不可不知的實用E-mail字彙

- **the cost is too high** 成本過高
- **the duration is too long** （施工）時間過長
- **not cost-effective** 不符成本效益
- **not economic-effective** 不符經濟效益
- **shortage of manpower** 人力短缺
- **shortage of funds** 資金短缺
- **not in urgent need of reconstruction** 沒有改建的迫切需要
- **not a high-priority problem** 非最優先需要解決的問題

04 更改方案內容

A Request Letter for Changes in Proposal

寫作要點 Key Points:

Step 1: 感謝對方提案。
Step 2: 提出提案中需要修改的問題。
Step 3: 請對方提交修改後的方案。

實際E-mail範例

寫信｜▼	刪除	回覆｜▼	寄件者：	Ginny

Dear Sir/Madam,

Introduction

In reply to your proposal that you sent to us last Monday, dated August 21st, 2017, we are pleased to inform you that it meets most of our requirements.

Body

However, you neglected to include the cost of tearing down the partition wall between two meeting rooms.

Closing

Could you please revise it with this specification and resubmit the proposal by the end of this week?

Thank you very much.

Best regards,
Signature of Sender
Sender's Name Printed

E-mail中譯

| 寫信▼ | 刪除 | 回覆▼ | 寄件者：Ginny |

您好：

開頭

此信目的是為了回覆您在上星期一，也就是 2017 年 8 月 21 日這天寄給我們的提案，我們很高興通知您，這份提案符合我們大部分的要求。

本體

然而，您忘記要把兩間會議室中隔間牆的拆除費用算進去了。

結尾

可以麻煩您修改這一個部分，並在這星期結束之前重新遞交提案嗎？
非常感謝您。

敬祝安康，
簽名檔／署名

各段落超實用句型

說明：畫底線部分的單字可按照個人情況自行替換

開頭 Introduction

❶ Thank you for submitting a proposal to create a website for our company.
謝謝您提出為本公司創設網站的提案。

❷ We have reviewed your proposal and had a discussion about it.
我們已經看過您的提案，並且討論過了。

❸ After reviewing your proposal, we find that the cost is way beyond our budget.
在看過您的提案之後，我們發現成本超過我們的預算太多了。

❹ I am writing to let you know that my team has reviewed your proposal and we are unfortunately conflicted over whether or not to accept.
我寫這封信是要告知您，我的團隊已經看過了您的提案，然而很遺憾的，我們目前很難決定究竟是否要接受。

本體 Body

❶ **I'm afraid that your proposal needs some revision.**
您的提案恐怕得做些修改。

❷ **We have made a change in our plans which requires some revision in your proposal.**
我們的計畫有個變動，因此需要您將提案做些修改。

❸ **I must ask you to revise your proposal as we decided to reduce our budget for the factory expansion.**
我必須請您修改您的提案，因為我們決定要降低擴充廠房的預算。

❹ **The repair and maintenance cost is over our budget.**
修繕維護成本超過我們的預算。

❺ **We would like to shorten the construction time.**
我們希望能縮短施工時間。

❻ **The cost of painting is neglected in your proposal.**
您的提案中忘記把油漆費用算進去了。

❼ **Please submit a revised proposal which reflects the changes of our plan.**
請提交一份反映我們計畫變動的修改後提案。

❽ **There are several things we liked about your proposal, but the amount of time needed is not one of them.**
您的提案有幾點我們相當喜歡，然而所需時間並不是我們喜歡的一點。

結尾 Closing

❶ **We look forward to the revised proposal.**
我們期待修改後的方案。

❷ **We hope to receive the modified proposal as soon as possible.**
我們希望能收到調整後的方案。

❸ **It would be great if you could have the proposal modified in accordance with our requirements by the end of March.**
如果您能在 3 月底前根據我們的要求調整提案內容，就太棒了。

❹ If necessary, we can arrange a personal meeting to discuss the specifications that need to be revised.

如有必要，我們可以安排一次會議，以討論需要修改的具體事項。

❺ Due to the changes, we are extending the deadline for your proposal to be received until <u>October 31st</u>.

因為修改的關係，我們將會延長提案提交截止日期至 <u>10 月 31 日</u>。

❻ Please let us know if it would work for you if we ask you to revise certain specifications.

請告知我們，若我們請您重新調整某些條件，對您而言是否方便。

 # 不可不知的實用E-mail字彙

- **neglect to include sth.** 忘記包含某事
- **extend the deadline** 延長截止日
- **meet the requirement** 符合要求
- **sth. needs modification** 需要作調整
- **revision required** 需要修改
- **extend the deadline** 延長截止時間

A Proposal Negotiation Letter

實際E-mail範例

寫信｜▼　刪除　回覆｜▼　　寄件者：　Ginny

Dear Sir/Madam,

Introduction

Thank you for your proposal for our restaurant expansion business plan.

Body

We have reviewed your proposal and we are quite happy with it. However, we found the estimated construction duration too long. According to our discussion with you dated June 30th, the construction is expected to be finished no later than November 30th, 2017. Therefore we must request you to evaluate the possibility of shortening the construction duration.

Closing

If you have any inquiries or questions, please send it by email no later this Friday.

Looking forward to hearing from you by July 27th.

Best regards,
Signature of Sender
Sender's Name Printed

E-mail中譯　寫信▼　刪除　回覆▼　寄件者：Ginny

您好：

> **開頭**

感謝您為本公司餐廳擴充營業計劃所做的提案。

> **本體**

我們已經看過您的提案，並且相當滿意。然而，我們認為預估施工時間太長。根據我們與您在 6 月 30 日所做的討論，我們希望工程能在 2017 年 11 月 30 日之前完成。因此我們必須請您評估縮短工程時間的可能性。

> **結尾**

如果您有任何問題，請在這週五之前以電子郵件寄給我們。

希望能在 7 月 27 日之前得到您的回覆。

敬祝安康，
簽名檔／署名

 ## 各段落超實用句型

說明：畫底線部分的單字可按照個人情況自行替換

開頭 Introduction

❶ **On behalf of our company, I would like to thank you for your time and effort that you have put into offering us this proposal.**

我要代表本公司感謝您為提供我們這份提案所付出的時間與努力

❷ **Thank you for submitting such a comprehensive proposal for our business plan.**

感謝您為我們的營業計劃，提交了一份如此全面的提案。

❸ Our team reviewed your proposal in detail, and overall we are quite happy with it.

我們團隊已經仔細看過您的提案,整體來說我們相當滿意。

❹ Our manager is actually trying to get in touch with <u>two</u> other suppliers, with a view to get a competitive pricing quoted.

我們經理其實正試著與其他<u>兩家</u>供應商聯繫,希望拿到很有競爭力的報價。

❺ I'm afraid this may be the issue that stops us from giving this contract to you.

恐怕這個問題會使我們無法跟貴公司合作。

❻ We are excited to have received your proposal, and would like to discuss a few details with you.

收到您的提案,我們非常興奮,希望能與您討論其中的一些條件。

本體 Body

❶ Our company reviewed your proposal and found the price too high.

我們公司看過您的提案,發現報價太高了。

❷ We would like to know if you could finish the project <u>one week</u> ahead of the schedule.

我們想知道您們是否可以比原定時間早<u>一星期</u>把案子做完。

❸ Our team has been given a specific budget for this project.

這項專案我們團隊所拿到的預算是固定的。

❹ We are expected to get everything ready for the beginning of operations of our new factory by <u>the end of this year</u>.

公司希望我們可以在<u>今年底</u>前將一切準備就緒,讓新廠房可以開始營運。

❺ We hope you understand that our budget for this project is limited.

希望您能理解,我們這項專案的預算有限。

❻ We are asked to get this contract within the budget.

我們被要求要在預算內拿到專案合約。

❼ We would have to look for other suppliers if your proposed price is not within our budget.

如果您的提案金額不能符合我們的預算,我們就必須開始找其他供應商。

❽ **Your price needs to be low than what you have quoted in your initial proposal.**

您的報價必須比您在一開始的提案中所報的還要低才行。

結尾 Closing

❶ **I hope to get your acceptance of this offer from our side.**

我希望貴公司能接受我方的提議。

❷ **We hope you can consider our offer and look forward to your positive reply.**

我們希望您能考慮我們的提議，並希望得到您正面的回覆。

❸ **Please do not hesitate to contact me if you have any questions or concerns. We will be glad to discuss further and find a way to meet the needs of both sides.**

若您有任何問題或疑慮，請不吝與我聯絡。我們很樂意與您進一步討論，找到符合雙方需求的方式。

 不可不知的實用E-mail字彙

- **proposed price** 提案報價
- **price negotiation** 價格協商
- **ask for a better price** 要求更便宜的價格
- **limited budget** 有限的預算
- **specific budget** 固定預算
- **get quotes from other suppliers** 取得其他供應商的報價

Part4
活動邀請篇

詢問出席聚會意願

An Event Invitation Letter

寫作要點 Key Points:

Step 1: 說明來信目的為邀請出席聚會。

Step 2: 清楚提供聚會時間與地點，並提供相關必要資訊。

Step 3: 請求對方給予回覆。

實際E-mail範例

| 寫信 ▼ | 刪除 | 回覆 ▼ | 寄件者： Ginny |

Dear Sir/Madam,

Introduction

I am writing this letter on behalf of my company to invite you to our 2017 Year-End Banquet.

Body

The banquet is scheduled to be held on December 23rd, 2017, 7:00 pm-9:00 pm at Regent Hotel. This is a major event of our company and it would be a great honor for us to have you at the banquet.

Closing

Please R.S.V.P. to let us know whether you will be able to attend.

We look forward to seeing you on December 23rd at 7 pm at Regent Hotel.

Best regards,

Signature of Sender

Sender's Name Printed

E-mail中譯

寫信 ▼　刪除　回覆 ▼　寄件者：Ginny

您好：

開頭

我謹代表本公司邀請您參加我們 2017 年的年終餐會。

本體

此次餐會預定於 2017 年 12 月 23 日晚上 7 點至 9 點，於晶華酒店舉行。這是本公司的重要活動，您的參與將是我們莫大的榮幸。

結尾

請回覆此信，好讓我們知道您是否可以出席。

期待 12 月 23 日下午 7 點能在晶華酒店見到您。

敬祝安康，
簽名檔／署名

各段落超實用句型

說明：畫底線部分的單字可按照個人情況自行替換

開頭 Introduction

❶ **On behalf of Mr. Smith, I would like to invite you and your family to the wedding of his daughter, Miss Lauren Smith's wedding party.**

我謹代表史密斯先生，邀請您與您家人參加他的千金──史密斯羅倫小姐的結婚派對。

❷ **It is with great pleasure that we invite you and your family to our celebration party.**

很高興能邀請您與您的家人來參加我們的慶功宴。

❸ **In celebration of the launch of our new products, we are giving a cocktail party and we'd like to invite you to join us.**

為了慶祝新品上市，我們將舉辦一個雞尾酒餐會，並希望能邀請您前來共襄盛舉。

❹ I am writing this letter to invite all of you to the newcomers' welcome party.

我寫這信是為了邀請各位參加新進同仁的歡迎會。

❺ We would like to take the opportunity to invite you to be our guest at our Christmas Party.

我們想利用這個機會邀請您擔任我們聖誕派對的嘉賓。

本體 Body

❶ It would be a great honor for us to have you at the party.

您的參與將會是我們莫大的榮幸。

❷ The charity dinner will start at 7 pm.

慈善餐會將會在晚上 7 點開始。

❸ The party will be held at the Grand Royal Hotel with a black tie dress code.

本派對將會在皇冠酒店舉行，需著正式服裝出席。

❹ The purpose of the party is to raise funds for charity and education.

本派對的目的為慈善及教育機構募款。

❺ We have also invited some popular singers to perform at the party for your entertainment.

我們也邀請了一些知名歌手到會場表演助興。

結尾 Closing

❶ We sincerely look forward to your esteemed attendance at the event.

我們誠摯地期望您能紆尊降貴，共襄盛舉。

❷ Please kindly send us a confirmation reply as soon as possible.

請您盡快給我們確認回覆。

❸ I look forward to the confirmation of your attendance.

期待能得到您出席的確認信。

❹ Look forward to seeing all of you at the welcome party.

期待能在歡迎會上見到各位。

❺ Please feel free to contact me if you have any questions or need any additional information.

如果您有任何問題或需要其他資訊，請儘管與我聯絡。

❻ If you would like any further information regarding the event, such as the dress code and parking availability, please do not hesitate to let me know.

若您需要更多關於此活動的進一步資訊，例如服裝規定及是否方便停車，請不吝告知我。

 不可不知的實用E-mail字彙

- **newcomer welcome party** 迎新會
- **farewell party** 送別會
- **celebration party** 慶功宴
- **cocktail party** 雞尾酒餐會
- **retirement party** 退休派對

- **tea party** 茶會
- **year-end party** 尾牙派對
- **annual banquet** 年終餐會
- **Christmas party** 聖誕派對
- **charity dinner** 慈善餐會

02 詢問出席會議意願

A Meeting Invitation Letter

實際E-mail範例

寫信 ▼　刪除　回覆 ▼　　寄件者：Ginny

Dear Sir/Madam,

Introduction

I am writing this letter to invite you to attend the Departmental Research Seminar of the month.

Body

The seminar will be held at Room 313 in the First Office Building at 9 am, August 20th, 2017.

Closing

Please let me know whether you will be attending the seminar by the end of this week, either by calling my extension number 2123 or emailing me at jennie123@jbs.com. Thank you very much.

Best regards,
Signature of Sender
Sender's Name Printed

E-mail中譯 　寫信｜▼　刪除　回覆｜▼　寄件者：Ginny

您好：

開頭

謹以此信邀請您出席本月份的部門研究研討會。

本體

研討會將會在 2017 年 8 月 20 日上午 9 點，假第一辦公大樓 313 室舉行。

結尾

無論您是否將出席本研討會，都請您透過分機 2123 或來信 jennie123@jbs.com 告訴我。非常感謝您。

敬祝安康，
簽名檔／署名

 # 各段落超實用句型

說明：畫底線部分的單字可按照個人情況自行替換

開頭 Introduction

❶ **This is to invite you to attend the weekly meeting.**

此信目的是要邀請您出席週會。

❷ **This letter serves to invite you to attend the monthly work report meeting.**

此信目的是要邀請您出席工作匯報月會。

❸ **I would like to invite you to participate in the Customer Service Skills Workshop.**

我想邀請您參加客服技巧專題討論會。

❹ **We would like to invite you to the product launch of our new lines of cosmetics.**

我們想邀請你參加本公司新系列化妝品的產品發表會。

❺ I am writing this letter to inquire about your inclination to attend our news conference.

我是寫信來詢問您是否有意願出席我們的記者招待會。

本體 Body ....

❶ The conference is scheduled at 2 pm on September 12th, 2017.

會議的時間定在 2017 年 9 月 12 日下午 2 點。

❷ The weekly work report meeting will be held in the Main Office Building at Rm. 1111.

本月工作匯報會議將會在總辦公大樓 1111 室舉行。

❸ The seminar is scheduled to be held on 31st August, 9:00 am at the meeting room.

該研討會預定於 8 月 31 日上午 9 點於會議室舉行。

❹ The news conference will be followed by the new product presentation by our Marketing team.

記者招待會將會在本公司行銷團隊負責的新品發表前舉行。

❺ The focus of the meeting will be centered on the sales programs and budgets.

會議的重點將會放在業務程序與預算上。

❻ There will be three key speakers at the event, each giving a talk for approximately 45 minutes.

此活動將有三名主講人，每人將演講約 45 分鐘。

結尾 Closing ....

❶ The agenda of the meeting is attached to this mail.

隨信附上會議的議程。

❷ Please see the attachment for the detailed agenda of the seminar.

詳細的研討會議程請參見附件。

❸ Please bring with you any ideas regarding teaching young children to the workshop.

請您在專題討論上提出任何有關幼兒教學的新點子。

❹ Please let me know whether you will attend the meeting or not A.S.A.P.

請盡快告訴我您是否會出席會議。

❺ I look forward to your prompt reply.

希望能立刻得到您的回覆。

❻ This is a friendly event and we would like to invite you to bring any snacks or drinks you would like to the meeting location.

這是個友善的活動,我們想邀您帶任何想要的點心或飲料到會議地點。

 # 不可不知的實用E-mail字彙

• **conference** 會議
• **meeting** 會議
• **seminar** 研討會(專題討論會)
• **workshop** 專題討論會;研討會
• **annual conference of ~** ～的年度會議
• **work report meeting** 工作報告會議
• **news conference** 記者招待會
• **new product presentation** 新品發表會
• **new product launch** 新品上市會
• **colloquium** 學術座談會

03 告知活動時間與流程

A Notice Letter of Meeting Agenda

寫作要點 Key Points:

Step 1: 說明來信目的為告知活動流程（說明活動名稱）。
Step 2: 簡短說明活動流程，或提供附件。
Step 3: 歡迎對方提出與活動流程相關的問題。

實際E-mail範例

寫信｜▼　刪除　回覆｜▼　寄件者：Ginny

Dear Sir/Madam,

Introduction

I am writing this letter with reference to the meeting scheduled on December 2nd, 2017.

Body

The agenda of the meeting is as follows.

13:00 – 13:30	Monthly Review
13:00 – 14:30	Departments' Work Reports
14:30 – 15:00	Tea Break
15:00 – 16:30	Year-end projections
16:30	Wrap up

Closing

If there's any specific issue that needs to be covered, please contact me by 5 pm today.

Best regards,
Signature of Sender
Sender's Name Printed

E-mail中譯　　寫信 ▼　　刪除　　回覆 ▼　　寄件者：Ginny

您好：

開頭

我撰寫此信是為了訂於 2017 年 12 月 2 日舉行的會議流程。

本體

本次會議流程如下：

13:00 – 13:30	本月檢討事項
13:00 – 14:30	各部室工作匯報
14:30 – 15:00	茶會
15:00 – 16:30	年終規劃事項
16:30	總結

結尾

如果需要在會議中加入任何具體議題，請在今天下午 5 點之前與我聯繫。

敬祝安康，
簽名檔／署名

 各段落超實用句型

說明：畫底線部分的單字可按照個人情況自行替換

開頭 Introduction

❶ **With reference to the upcoming annual meeting scheduled at 2 PM on June 10th, 2017,...**

有關 2017 年 6 月 10 日下午 2 點舉行，即將到來的年度會議，……

❷ **This letter is to inform you of the agenda of our upcoming monthly work report meeting.**

這封信是要通知您即將到來的每月工作報告會議的流程。

❸ With this email I am sending you the agenda of the meeting scheduled at 10 am tomorrow morning.

我以這封信寄給各位排定於明天上午 10 點的會議流程。

❹ The agenda for the teaching workshop is now available.

教學研討會的流程已經出來了。

❺ I send you herewith the agenda for the October 16th Business Conference.

我隨函附上 10 月 16 日商務會議的流程。

本體 Body

❶ The agenda of the conference is listed below.

會議流程詳列如下。

❷ You can find the detailed agenda of the seminar in the attachment.

您可以在附件中看到詳細的研討會流程。

❸ Attached is the agenda of the news conference.

附件為記者招待會的流程。

❹ Pleased find the attached the documents for this month's assembly.

本月會議文件請見附件。

❺ The meeting will commence at 9 am and will conclude approximately at 12 pm.

會議將會在上午 9 點開始，並大約在中午 12 點結束。

❻ Please note that the meeting will start at 2 pm sharp. We would highly appreciate it if you arrive on time.

請注意，會議將在下午 2 點整開始，若您準時抵達，我們將感激不盡。

結尾 Closing

❶ Please let me know if there are any particular subjects that need to be added.

如果有任何需要加入流程的特別議題，請讓我知道。

❷ Please contact me should you have any questions regarding the assembly.

如果有任何關於會議的問題請與我聯絡。

❸ **Please be informed that there will not be a break during the meeting.**
請注意，會議期間將不會有休息時間。

❹ **Your presence is requested for the assembly.**
請務必出席會議。

❺ **Please be advised that everyone is requested to come on time.**
請注意，所有人都必須準時出席。

❻ **If you have any questions or may have trouble attending, please contact <u>Mr. Lin</u> at <u>(02)23456789</u>.**
如果您有任何問題或可能無法參加，請撥電話至 <u>(02)23456789</u> 聯絡<u>林先生</u>。

 不可不知的實用E-mail字彙

- **agenda items** 流程項目
- **conference agenda** 會議流程
- **seminar agenda** 研討會流程
- **welcome/ introduction** 歡迎詞／引言
- **opening remarks** 開場詞

- **Committee Leader Remarks** 委員會長致辭
- **matters arising** 前議事項
- **close of meeting** 散會

A Notice Letter for Change of Meeting Time/Agenda

寫作要點 Key Points:

Step 1: 說明活動時間或流程有所更動。
Step 2: 清楚表示更改的內容。
Step 3: 感謝對方留意變更。

實際E-mail範例

| 寫信 ▼ | 刪除 | 回覆 ▼ | 寄件者： Ginny |

Dear Sir/Madam,

Introduction

I am writing this email with regard to the teaching workshop scheduled to be held at 9 am this Wednes day.

Body

Please note that the meeting time and location have been changed.

The meeting time has been changed from 9 am, June 28th to 9 am, June 30th,

and the meeting location has been changed from Conference Room 011 to Conference Room 020 in order to accommodate a larger group of attendees.

Closing

The meeting agenda remains unchanged.

Look forward to seeing you all on June 30 at Conference Room 020.

Best regards,
Signature of Sender
Sender's Name Printed

E-mail中譯

寫信 ▼　刪除　回覆 ▼　寄件者： Ginny

您好：

開頭

這封信是有關原訂於本週三上午 9 點舉行的教學研討會。

本體

請注意會議時間以及會議地點有所變動。

會議時間已從 6 月 28 日上午 9 點改至 6 月 30 日上午 9 點。

而會議地點已經從 011 會議室移至 020 會議室以容納更多的與會者。

結尾

會議流程維持不變。

期待 6 月 30 日在 020 會議室見到各位。

敬祝安康，
簽名檔／署名

 ## 各段落超實用句型

說明：畫底線部分的單字可按照個人情況自行替換

開頭 Introduction

❶ **This letter serves to officially inform you of the change of the meeting date.**

這封信的目的是要正式通知您會議的日期變更。

❷ **I am writing this letter to inform you that the date of the meeting has been changed.**

謹以此信通知您，會議時間已經改了。

❸ **Please be informed that we have modified the meeting agenda.**

請注意，我們已經修改了會議流程。

❹ By means of this letter I would like to inform you of the change of the meeting agenda.

藉由這封信，我想通知各位有關會議流程變動一事。

❺ Please kindly be advised that the agenda of the meeting has been changed.

請您留意會議流程有所變更。

本體 Body

❶ The date of the business conference has been changed from May 12th to May 19th.

商務會議日期已經從 5 月 12 日改至 5 月 19 日。

❷ Due to the limited space of the room, we have decided to change the meeting venue.

由於會議室空間有限，我們已經決定更改會議舉行地點。

❸ Monday's conference has been changed from Room 101 to Room 403 in order to accommodate all the conferees.

為了容納所有與會者，週一的會議已經從 101 室改到 403 室。

❹ Due to time constraints, we have to defer the project presentation to a future meeting.

由於時間有限的關係，我們必須將專案簡報延遲至將來的會議中。

❺ In terms of the agenda, we decided to run a workshop after the lecture.

在流程方面，我們決定在演講結束後，舉辦一個專題討論會。

❻ Unfortunately, we have to postpone the date of the conference from the 15th to the 16th, though the location remains the same.

很遺憾地，我們必須把研討會的日期從 15 號延至 16 號，不過地點還是一樣。

結尾 Closing

❶ Please see the attachment for the latest agenda.

最新的流程表請見附件。

❷ Thank you for your attention in this matter.

謝謝您對此事的注意。

❸ I can be reached at extension <u>2023</u> with any questions you may have.

有任何問題都可以撥分機 <u>2023</u> 與我聯絡。

❹ Please contact me directly if you have any concerns with the change of agenda.

如果有任何關於流程異動的問題，請直接與我聯絡。

❺ Should you require any additional information, please feel free to call me at extension <u>2023</u>.

如果您需要任何其他資訊，請儘管撥打分機 <u>2023</u> 給我。

 # 不可不知的實用E-mail字彙

- **notice of time change**　時間異動通知
- **notice of location change**　地點異動通知
- **notice of agenda change**　流程異動通知
- **time constraint**　時間限制
- **limited space of the room**　地點空間有限
- **add an item to the agenda**　將一個項目加入流程
- **remove an item from the agenda**　將一個項目從流程中移除

05 告知活動取消

A Notice of Cancellation Letter

實際E-mail範例

| 寫信 ▼ | 刪除 | 回覆 ▼ | 寄件者： Ginny |

Dear Sir/Madam,

Introduction

We regret to announce that the "Telephone Marketing Skills Workshop", scheulded for July 24th 2017 at JBS Headquateres in Taipei, has been cancelled.

Body

The reason of the cancellation is that our main presenter, Mr. Stevenson, has to attend an important meeting in Tokyo. He has no option but to cancel the workshop on an emergency basis.

Closing

We apologize for any inconvenience the cancellation may have caused.

We will reschedule the workshop as soon as Mr. Stevenson returns.

Thank you for your understanding.

Best regards,
Signature of Sender
Sender's Name Printed

E-mail中譯　　寫信 ▼　　刪除　　回覆 ▼　　寄件者：Ginny

您好：

開頭

我們很遺憾地宣布，定於 2017 年 7 月 24 日在 JBS 臺北總公司舉行的「電話行銷技巧研討會」已經取消。

本體

取消原因是我們的主講人──史蒂文森先生，必須參加一場在東京舉行的重要會議。由於事出緊急，他不得不取消這次研討會。

結尾

我們對於這次取消活動可能造成的不便，向您致歉。

待史蒂文森先生回國，我們就會立刻重新安排研討會的時間。

感謝您的體諒。

敬祝安康，
簽名檔／署名

各段落超實用句型

說明：畫底線部分的單字可按照個人情況自行替換

開頭 Introduction

❶ **I am writing this letter to inform you that the meeting scheduled on November 11ᵗʰ has been cancelled.**

謹以此信通知您，原定於 11 月 11 日的會議已經取消。

❷ **This letter serves to inform you of the cancellation of the meeting scheduled to be held at 11 am on November 11ᵗʰ.**

這封信是要通知您原定於 11 月 11 日上午 11 點舉行的會議業已取消。

❸ **The purpose of this letter is to notify you that the meeting is cancelled.**

這封信的目的是要通知您會議取消了。

❹ We with apology would like to inform you that our <u>new product launch presentation</u> has been cancelled.

很抱歉通知您，我們的<u>新產品上市發表會</u>已經取消了。

❺ Please be informed that <u>tomorrow's luncheon</u> is cancelled.

請留意<u>明天的午餐會議</u>已經取消了。

本體 Body

❶ The reason for cancellation is that <u>the meeting has a time conflict with another meeting</u>.

會議取消的原因是<u>因為與另一場會議時間重疊</u>。

❷ We have to cancel the lecture because <u>Professor Wood has postponed his trip to Taipei</u>.

我們必須取消這次講座，是因為<u>伍德教授延遲了他到臺北的行程</u>。

❸ We decided to cancel the <u>annual staff dinner because we have difficulty finding an ideal location</u>.

我們決定取消<u>員工聚餐</u>，是<u>因為我們找不到理想的場地</u>。

❹ We have no option but to cancel the <u>annual company trip</u> due to <u>lack of funds</u>.

因為<u>經費有限</u>，我們不得不取消<u>年度員工旅遊</u>。

❺ The meeting is cancelled due to <u>the typhoon day-off</u>.

因為<u>颱風假</u>之故，會議取消。

❻ We are sorry to inform you that the <u>talk</u> has been cancelled due to <u>a sudden emergency in the speaker's family</u>.

很抱歉必須告知您，由於<u>講者家中的緊急狀況</u>，演講已取消。

結尾 Closing

❶ I will let you know if the meeting is rescheduled.

如果會議改期，我會通知各位。

❷ We really apologize for the cancellation.

謹對（活動／會議）取消一事，向您致歉。

❸ We are sorry for any inconvenience it has caused.

造成任何不便，我們深感抱歉。

❹ **The lecture will be rescheduled as soon as possible.**
　我們會儘快重新排定講座時間。

❺ **We will reschedule the party once we find the ideal location.**
　一旦找到理想場地，我們將重新安排派對時間。

 不可不知的實用E-mail字彙

- **staff dinner** 員工聚餐
- **annual staff dinner** 年度員工聚餐
- **department dinner** 部門聚餐
- **office party** 公司聚餐
- **company trip** 員工旅遊
- **lunch meeting** 午餐會議
- **appreciation luncheon** 感恩午餐會
- **teacher appreciation banquet** 謝師宴
- **skills training lecture** 技能訓練講座
- **employee training course** 員工訓練課程
- **speech** 演講
- **lecture** 講座

06 答應他人邀請

An Invitation Acceptance Letter

實際E-mail範例

| 寫信 ▼ | 刪除 | 回覆 ▼ | 寄件者： Ginny |

Dear Sir/Madam,

Introduction

Thank you for your kind invitation to your new product launch presentation on January 12th.

Body

I have been a loyal customer of your products for the past ten years and always look forward to your new products.

I am very pleased to attend the new product launch presentation.

Closing

I appreciate your thinking of me.

Look forward to the event!

Best regards,
Signature of Sender
Sender's Name Printed

E-mail中譯

寫信 ▼　　刪除　　回覆 ▼　　寄件者：Ginny

您好：

開頭

感謝您邀請我參加貴公司 1 月 12 日的新產品上市發表會。

本體

過去十年來，我一直是貴公司產品的忠實顧客，而且總是對你們的新產品十分期待。

我很樂意出席您的新品上市發表會。

結尾

感謝您想到我。

很期待能參加活動！

敬祝安康，
簽名檔／署名

各段落超實用句型

說明：畫底線部分的單字可按照個人情況自行替換

開頭 Introduction

❶ **Thank you for inviting me to the opening ceremony of your new branch office.**

感謝您邀請我參加您的新分公司之開幕典禮。

❷ **I appreciate your invitation to the groundbreaking ceremony of your new factory.**

感謝您邀請我參加貴公司新廠動土典禮。

❸ **I would like to thank you for inviting me to the ~ Conference.**

謝謝您邀請我參加～的會議。

❹ **I was delighted to receive the invitation to your daughter's wedding.**

非常開心能收到貴千金婚禮的邀請函。

❺ In response to your invitation to the Career Fair 2017, ...

此為回覆您邀請本公司參加 2017 年度就業博覽會的邀請函，……

❻ I was very excited and flattered to have been invited to your company's year-end banquet.

能受邀至您公司的尾牙餐會，我非常興奮，受寵若驚。

本體 Body

❶ I am pleased to join your function.

很高興能參與您的盛會。

❷ We are delighted to accept the invitation.

我們很開心能接受邀請。

❸ It will be an honor to attend your luncheon.

能參加您的午餐會是我的榮幸。

❹ I would like to confirm my attendance to your annual staff dinner.

我要跟您確認我將出席貴公司的年度員工餐會。

❺ It is with great pleasure that I accept the invitation to join your opening ceremony.

很榮幸地，我接受參加貴公司開幕典禮的邀請。

❻ I would, of course, love to be present, and I'm very grateful to you for thinking of me.

我當然樂意參加，也很感激您想到我。

結尾 Closing

❶ We look forward to the event.

我們對此活動倍感期待。

❷ I look forward to a great evening with you.

我期待能與您共度一個美好的夜晚。

❸ It would be great if you could send me a map of the meeting location.

如果您能寄給我一張會議地點的地圖，就太好了。

❹ **Please let me know if there's a dress code.**
如果有服裝規定，請告訴我。

❺ **Please do keep me informed if anything needs to be prepared prior to the event.**
如果有任何需要在活動前準備好的事項，請務必通知我。

❻ **I look forward to having a great night with you and your team. Please do not hesitate to let me know if there is anything I need to bring or any preparations to make.**
我很期待與您和您的團隊度過美好的一夜。若我需要帶什麼東西或做什麼準備，請不吝讓我知道。

 # 不可不知的實用E-mail字彙

（常見的服裝規定）
- **casual** 便裝
- **business casual** 商務便裝
- **smart casual** 正式休閒裝
- **business/informal** 商務裝（非正式）
- **formal wear** 正式服裝
- **black tie** 黑領結（指男性需要戴黑領結）
- **evening gown** 晚禮服（指女性需要穿晚禮服）

07 婉拒他人邀請

An Invitation Declination Letter

寫作要點 Key Points:

Step 1: 感謝對方邀請出席活動。
Step 2: 婉拒邀請，並提出不得不拒絕邀請的原因。
Step 3: 祝福對方活動順利成功。

實際E-mail範例

寫信 ▼　　刪除　　回覆 ▼　　寄件者： Ginny

Dear Sir/Madam,

Introduction

Thank you very much for inviting me to attend your annual appreciation luncheon with you and your distinguished guests.

Body

While I would be delighted to join you, I will have to attend a meeting of my professional organization on September 18th. It is with regret that I will have to miss the luncheon with you.

Closing

Best wishes for an enjoyable and successful event.

Best regards,
Signature of Sender
Sender's Name Printed

E-mail中譯

寫信 ▾　刪除　回覆 ▾　寄件者：Ginny

您好：

開頭

非常感謝您邀請我，連同您與您的貴賓一同出席貴公司一年一度的感恩餐會。

本體

雖然我很想參加，但是我在 9 月 18 日這天必須參加公司內部的一場會議。很遺憾我必須錯過與您的餐會。。

結尾

祝您活動愉快且成功。

敬祝安康，
簽名檔／署名

各段落超實用句型

說明：畫底線部分的單字可按照個人情況自行替換

開頭 Introduction

❶ Thank you for inviting me to your <u>opening ceremony</u>.

感謝您邀請我參加您的<u>開幕典禮</u>。

❷ Thank you for including me on your guest list for the upcoming <u>year-end banquet</u>.

感謝您將我列在貴公司即將到來的<u>尾牙餐會</u>賓客名單上。

❸ I am flattered to be invited to attend such a function.

能受邀參加如此盛會，我感到受寵若驚。

❹ I am pleased and honored to receive your invitation to your <u>opening ceremony</u>.

收到您開幕典禮的邀請函，我感到十分高興與榮幸。

❺ It is a great honor for me to be invited to your <u>ground-breaking ceremony</u>.

能獲邀參加貴公司的<u>動土典禮</u>，本人備感光榮。

❻ I was thrilled to receive your invitation to your company's <u>get-together</u>.

獲邀至您公司的<u>聚會</u>，我非常興奮。

本體 Body

❶ Unfortunately I will not be able to attend.

很可惜我將無法出席。

❷ I will not be able to attend because of <u>a prior engagement</u>.

因為有<u>其他優先安排</u>，我將無法出席。

❸ It is with regret that I must decline.

很遺憾的是，我必須婉拒您的邀請。

❹ I will <u>not be in the country</u> on the day of your function.

您的盛會舉辦當天，<u>我人不在國內</u>。

❺ We will not be able to attend the charity party due to <u>scheduling constraints</u>.

<u>受限於行程安排</u>，我們將無法出席這次慈善晚會。

❻ Unfortunately, while I would really love to be there, my schedule simply does not permit me to do so.

很遺憾地，雖然我真的很希望能夠參與，我的行程實在不允許我參加。

結尾 Closing

❶ Thank you for thinking of me.

感謝您們想到我。

❷ Best wishes for a successful event.

祝您有個成功的活動。

❸ We hope to be able to attend next time.

我們希望下次能夠參加。

❹ **I'll be there in spirit.**
我的精神與你們同在。

❺ **Wish you a wonderful <u>celebration party</u>.**
祝您有個完美的<u>慶功宴</u>。

❻ **Thank you so much for the invitation nevertheless, and I look forward to getting together with you all some other time in the future.**
還是感謝您的邀請，我很期待能與你們大家在未來某個時間相聚。

 # 不可不知的實用E-mail字彙

- **prior engagement** 更優先（重要）的事情
- **scheduling constraint** 行程安排限制
- **budgetary constraint** 預算限制
- **time conflict** 時間衝突
- **business commitment** 公事安排

Part5
訂單交涉篇

01 要求廠商估價

A Quotation Request Letter

寫作要點 Key Points:

Step 1: 請求對方對某服務或產品提供報價。
Step 2: 請求報價，或提供對方報價所需相關資訊。
Step 3: 請對方在具體時間內回覆報價。

實際E-mail範例

寫信 ▼	刪除	回覆 ▼	寄件者： Ginny

Dear Sir/Madam,

Introduction

I am writing this letter for a price quote on replacing the central air conditioning system in our office.

Body

The footage of our office is approximately 6,840 square feet with three individual central air conditioning units. It is our hope that the work can be done in less than one week, so please take that into account in your price quote.

Closing

Please provide us with pricing information and the estimated length of work you need to complete the job.

We look forward to hearing from you as soon as possible.

Best regards,
Signature of Sender
Sender's Name Printed

E-mail中譯　寫信 ▼　刪除　回覆 ▼　寄件者：Ginny

您好：

開頭

謹以此信請您提供更換本公司辦公室中央空調系統的報價。

本體

我們辦公室的面積大約是 6,840 平方尺，並且有三個獨立中央空調機。我們希望能在一週內完工，所以報價時請把這點考慮進去。

結尾

請提供我們報價，以及您估計完工所需的時間。

我們期待能盡快得到您的回覆。

敬祝安康，
簽名檔／署名

各段落超實用句型

說明：畫底線部分的單字可按照個人情況自行替換

開頭 Introduction

❶ **We are interested in purchasing your Color Laserjet Pro MFP 3300 for our office use.**

我們有興趣採購貴公司生產的彩色雷射專業級印表機，型號 MFP3300，作為辦公用途。

❷ **I am writing this letter to inquire about the price of your multi-function printer MFP 3300.**

我寫這封信是為了詢問貴公司多功能印表機 MFP3300 的價格。

❸ **We are currently planning to refurbish the office of our Taipei Branch.**

我們目前計劃要翻新我們臺北分公司的辦公室。

❹ **We would like to request a quotation on the following items.**

我們想要請你們提供下列產品的報價。

❺ We would be very grateful to you if you could provide us the quotation of your lift repair and maintenance service.

如果您能提供我們電梯維修服務的報價，我們將非常感激。

❻ I'm writing to ask for a quotation on the office file cabinets your company provides.

我寫這封信是希望能夠得到您公司的辦公室檔案櫃產品的報價。

本體 Body

❶ We are in urgent need of 1,000 multi-function printers.

我們急需一千台多功能事務機。

❷ The items we need are listed as below.

我們所需要的品項詳列如下。

❸ We would like to request you to send us your quotation at the earliest.

我們想請您盡快將報價單寄給我們。

❹ Please include packing and delivery in your quoted prices.

請將包裝及運送費用包含在您的報價中。

❺ Please send us your sales quotation with terms of payment.

請將付款條件連同產品報價一同寄給我們。

❻ We would highly appreciate a list of the prices for each item as well as the available payment methods.

若能獲得各產品的價格清單及可用的付款方式，我們將感激不盡。

結尾 Closing

❶ Your prompt reply would be highly appreciated.

若您能盡快回覆，我們將不勝感激。

❷ We would appreciate your quotation by the end of the week for our consideration.

希望您能在這週結束之前提供報價單讓我們考慮，感激不盡。

❸ We hope to receive your quotation no later than this Friday.

我們希望能在本週五之前收到您的報價單。

❹ Please send us your quotation by <u>August 18th, 2017</u> for our acquisition consideration.

請在 <u>2017 年 8 月 18 日</u>之前將您的報價單寄給我們，以便我們考慮採購。

❺ Provided the prices meet with our budget, we would place our order within <u>a week</u>.

假如價格符合我們的預算，我們將會在<u>一週</u>內下單。

❻ As the renovation process will be starting soon, we would be very grateful if we could receive your quotation before <u>the end of this week</u>.

由於快要開始重新裝潢了，我們非常希望能夠在<u>這週結束前</u>獲得您的報價。

 ## 不可不知的實用E-mail字彙

- **request for quotation** 請求報價單
- **provide the quotation** 提供報價單
- **meet with designated budget** 符合既定預算
- **acquisition consideration** 採購考慮
- **service charge** 服務費
- **packing charge** 包裝費
- **delivery charge** 運費

A Letter of Quotation

寫作要點 Key Points:

Step 1: 感謝對方對公司產品有興趣。
Step 2: 向對方提供估價。
Step 3: 表示希望能有合作機會。

 實際E-mail範例

| 寫信 ▼ | 刪除 | 回覆 ▼ | 寄件者： Ginny |

Dear Sir/Madam,

Introduction

In reply to your enquiry about the price of repairing and maintaining the elevators throughout your office building, I am pleased to provide the quotation as attached.

Body

In view that we haven't had the pleasure of working with you before, we would like to give a one-year discount of NT$5,000 per month, providing you agree to the terms within the next seven days.

Closing

We look forward to working with you, and hope this is the beginning of our long-term relationship.

Best regards,
Signature of Sender
Sender's Name Printed

E-mail中譯

寫信｜▼　刪除　回覆｜▼　寄件者：Ginny

您好：

開頭

我很樂意回覆您詢問貴公司整棟辦公大樓電梯修理維護費用一事並提供報價。報價單如附件。

本體

由於我們過去從未和您合作，若您在未來七天內同意合作條款，我們將提供您為期一年，每個月五千元的優惠折扣。

結尾

我們期待能與貴公司合作，並希望這次是我們長期合作的開端。

敬祝安康，
簽名檔／署名

各段落超實用句型

說明：畫底線部分的單字可按照個人情況自行替換

開頭 Introduction

❶ **Thank you for your enquiry for a quotation.**

感謝您詢問報價。

❷ **Thank you for inquiring about our lift repair and maintenance service.**

感謝您詢問本公司的電梯維修服務費用。

❸ **Thank you for being interested in our latest multi-function dishwasher, DW101.**

感謝您對本公司最新的多功能洗碗機 DW101 有興趣。

❹ **We would like to thank you for your letter dated April 27th, 2017, inquiring about our central air conditioning system.**

感謝您於 2017 年 4 月 27 日來信詢問有關本公司的中央空調系統。

❺ In reply to your quotation request letter dated March 31st, 2017, ...

此信為回覆您 2017 年 3 月 31 日請求報價的來信，……

❻ We are grateful for your interest in our products and will be more than happy to provide you a list of prices.

我們非常感激您對我們產品感興趣，也很樂意提供您價格清單。

本體 Body

❶ We are pleased to provide the quotation you requested.

我們很樂意提供您所要求的報價單。

❷ Please find the attached quotation for the service you enquired.

您所詢問的服務項目報價，請見附件。

❸ Please note that the service charge is excluded in the quoted rates.

請注意，所報價格並不包含服務費。

❹ Please be advised that this quotation is valid for 14 days from the date shown above.

請您留意此報價金額僅在本文件所示日期十四日內有效。

❺ The quotation you requested is as follows.

您所要求的報價如下。

❻ I have attached the quotation for the items you need. Please note that the prices listed are only valid for this quarter.

我已經附上了您所需要的物品報價。請注意，此處列出的價格只適用於這一季。

結尾 Closing

❶ If you need any further details to meet your requirements, please feel free to contact me.

如果您希望進一步了解任何符合您要求的細節，請儘管與我聯繫。

❷ I hope our offer meets with your approval.

我希望我們的報價能獲得您的認同。

❸ Should you require any assistance with this quotation, please do not hesitate to contact us.

如果您需要任何與這份報價單有關的協助，請直接與我們聯絡。

❹ **Thank you for your interest.**
感謝您對本公司產品的興趣。

❺ **Looking forward to working with you.**
期待與您合作。

❻ **If you have any questions regarding the quotation or require further details, feel free to let me know.**
若您對此報價有什麼疑問，或需要更多細節，請不吝告知我。

 ## 不可不知的實用E-mail字彙

- **new customer offer** 新客戶優惠
- **first-year discount** 首年優惠
- **special discount** 特別優惠
- **discount on all items** 全面優惠
- **discount for existing customers** 舊客戶優惠
- **agree to terms** 同意條款

A Quotation Negotiation Letter

寫作要點 Key Points:

Step 1: 表示收到對方報價。
Step 2: 向對方提出議價要求。
Step 3: 請求對方回覆，並表達合作意願。

實際E-mail範例

寫信 ▼	刪除	回覆 ▼	寄件者： Ginny

Dear Sir/Madam,

Introduction

We have received the quotation from you. Thank you very much.

Body

We have reviewed your quotation and did not find it as competitive as we expected.

We would very much like to work with you as your products fulfill our requirements; however, we would have to seek for other quotes from other suppliers if we cannot get your quote within our budget.

Closing

We hope to continue working with you. Could you give us a new price quotation no later than February 12th?

Thank you very much.

Best regards,
Signature of Sender
Sender's Name Printed

E-mail中譯

| 寫信 ▼ | 刪除 | 回覆 ▼ | 寄件者：Ginny |

您好：

開頭

我們已經收到您的報價單。非常感謝您。

本體

我們看了您的報價單，發現價格不如我們所預期的具競爭力。

我們非常想要跟你們合作，因為貴公司的產品符合我們的要求；然而，如果報價超出預算，我們就必須尋求其他廠商的報價。

結尾

我們希望能夠繼續與你們合作，所以麻煩您在 2 月 12 日之前給我們一份新的報價單好嗎？

非常感謝您。

敬祝安康，
簽名檔／署名

各段落超實用句型

說明：畫底線部分的單字可按照個人情況自行替換

開頭 Introduction

❶ **Thank you for your quote.**
謝謝您的報價。

❷ **Thank you for submitting the quotation we requested.**
謝謝您提交我們所要求的報價單。

❸ **Thank you for the quote you provided on April 20th, 2017.**
感謝您在 2017 年 4 月 20 日提供的報價單。

❹ **I appreciate your quotation letter dated May 10th, 2017, for installing a solar water heating system in our factory.**
感謝您在 2017 年 5 月 10 日提供安裝本工廠太陽能熱水器的報價。

⑤ Regarding the quotation you provided <u>this past Monday</u>, ...

有關於您在這星期一提供給我們的報價單，……

⑥ I am very grateful for your prompt reply with the price list for your services.

我很感激您立即將您公司各項服務的價格清單回覆給我。

本體 Body

❶ We have looked into the quote and find that it doesn't quite meet with our designated budget.

我們已經看過您的報價單，並發現它不太符合我們的既定預算。

❷ I hope you understand that we are given a specific budget for this quote.

我希望您能明白，我們在這份報價上的預算有限。

❸ I sincerely hope we can get your quote within our budget.

我誠摯希望你們能提供我們預算內的報價。

❹ We need your support to reduce our costs.

我們需要您支持我們降低成本。

❺ I will have to seek quotes from other suppliers if we cannot get your quote within our budget.

如果你們無法提供我們預算內的報價，我就必須尋求其他廠商的報價了。

❻ As much as we would love to purchase these items from you, our budget simply does not permit it.

雖然我們很想和你們購買這些物品，我們的預算實在不允許我們這麼做。

結尾 Closing

❶ I am waiting for a rapid reply from your end.

期待貴公司儘速回覆。

❷ We hope to continue working with you.

我們希望能與您繼續合作。

❸ Please submit a new quotation as requested.

請依要求提交一份新的報價單。

❹ Please resubmit your quotation by <u>the end of the month</u>.
請在<u>本月底</u>前重新提交您的報價單。

❺ We would like to receive your new quotation as soon as possible.
我們希望能盡快收到您的新報價單。

❻ If possible, we would appreciate it if you reconsider the offer and send us a revised quotation.
如果可能的話,希望您能夠重新考慮開價,再寄送一份更改過的報價給我們。

 # 不可不知的實用E-mail字彙

- **resubmit a quotation**　重新遞交報價單
- **ask for a reduced price**　要求降價
- **negotiate on sales price**　議價
- **competitive quotation**　具競爭力的報價
- **obtain a fair and reasonable price**　取得公平合理價格
- **get the lowest price**　取得最低價格
- **create a long-term relationship**　建立長期關係

A Quotation Acceptance Letter

寫作要點 Key Points:

Step 1: 感謝對方報價。
Step 2: 表示對對方報價感到滿意，並決定下單。
Step 3: 表達希望雙方合作愉快。

 實際E-mail範例

| 寫信 ▼ | 刪除 | 回覆 ▼ | 寄件者： Ginny |

Dear Sir/Madam,

Introduction

This letter serves to inform you that we have received your quotation letter dated March 15th, 2017. Thank you for the time and effort you have put into providing us the quotation.

Body

We are very pleased with your quotation as it meets our designated budget. We are also very impressed with your confidence in completing the job on schedule. Therefore, we would like to accept your services.

Closing

We would like you to proceed with the preparation work, so that you are able to commence the work as soon as the cooperation agreement is signed by both parties.

Best regards,
Signature of Sender
Sender's Name Printed

E-mail中譯

寫信 ▼　刪除　回覆 ▼　寄件者：Ginny

您好：

開頭

特以此信通知您，我們已經收到您於 2017 年 3 月 15 日寄來的報價單。感謝您為提供我們此報價單所付出的時間與努力。

本體

由於您的報價符合我們既定預算，因此我們對您的報價非常滿意。我們也非常欣賞您有如期完工的自信。因此我們決定接受您的服務。

結尾

我們希望你們能著手進行準備工作，如此一來一旦雙方簽署合作協議之後，你們就可以馬上開始施工。

敬祝安康，
簽名檔／署名

各段落超實用句型

說明：畫底線部分的單字可按照個人情況自行替換

開頭 Introduction

❶ Thank you very much for your quotation.
非常感謝您的報價。

❷ Thank you for the quote you provided us on November 23rd, 2017.
感謝您在 2017 年 11 月 23 日提供給我們的報價。

❸ Thank you for taking the time to provide us a quote for replacing the central air conditioning system in our office.
感謝您撥冗提供本公司辦公室中央空調系統更換之報價。

❹ Thank you for sending us the quotation.
感謝您寄報價單給我們。

❺ With reference to the quotation you provided us on August 10th, 2017, ...

有關您在 2017 年 8 月 10 日所提供給我們的報價單，……

❻ This is regarding the quotation you sent us yesterday morning, on October 4th.

這封信是關於您昨天（10 月 4 日）早上寄送給我們的報價。

本體 Body

❶ Your quote meets our designated budget.

您的報價符合本公司的既定預算。

❷ It is with great pleasure that I'm informing you that we have decided to accept your services.

很高興通知您，我們已經決定接受貴公司的服務。

❸ We would like to accept your quote.

我們要接受您的報價。

❹ Please let us know when you can start working.

請告訴我們你們何時可以開始施工。

❺ Please proceed with the preparation work, so as to commence the work as soon as the agreement is signed.

請著手準備工作，一旦簽好合約你們就能開工。

❻ We are grateful for your speedy reply, and would like to accept your services.

我們很感激您的快速回覆，也很樂意接受您的服務。

結尾 Closing

❶ We will send you our order by fax today.

我們會在今天以傳真方式將我們的訂單傳送給您。

❷ Attached is the contract that needs to be signed by both parties.

附件為須由雙方簽署之合約。

❸ Please sign two copies of the contract and return by December 12th, 2017.

請簽署合約一式兩份，並於 2017 年 12 月 12 日寄回。

❹ **Please review the terms and conditions of the contract.**
請詳閱合約內容及條款。

❺ **Thank you for your continued interest in doing business with us.**
感謝您持續與本公司合作。

❻ **Attached is the contract we have drafted. If you have any questions regarding the details, do let us know.**
我們附上了擬好的合約。若您對細節有什麼疑問，請告知我們。

 # 不可不知的實用E-mail字彙

- **order** 訂單
- **place order** 下單
- **contract** 合約
- **two copies of the contract** 一式兩份之合約
- **accept the quotation** 接受報價
- **terms and conditions** 條約及條件
- **terms of payment** 付款條件

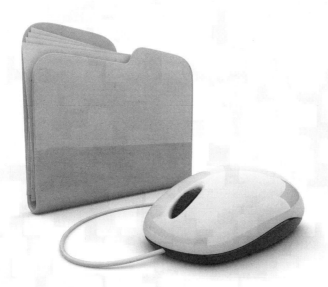

05 拒絕廠商的報價

A Quotation Declination Letter

寫作要點 Key Points:

Step 1: 感謝對方提供報價。
Step 2: 表示無法接受對方報價。
Step 3: 表示希望未來仍有合作機會。

 實際E-mail範例

| 寫信▼ | 刪除 | 回覆▼ | 寄件者： | Ginny |

Dear Sir/Madam,

Introduction

Thank you for submitting your quotation of the painting work for our new office building.

Body

It is with great regret that I have to inform you that we are unable to accept your quotation. We have decided to accept a competitor's offer because the quoted price was slightly better.

Closing

We hope there are other opportunities to do business with you in the future.

Best regards,
Signature of Sender
Sender's Name Printed

E-mail中譯　[寫信｜▼]　[刪除]　[回覆｜▼]　寄件者：[Ginny]

您好：

開頭

感謝您提供為本公司新辦公大樓進行粉刷工程的報價。

本體

非常遺憾通知您，我們無法接受您的報價。

我們已經決定接受一家競爭廠商的報價，因為他們的報價稍微便宜了一些。

結尾

希望將來還有其他機會可以與你們合作。

敬祝安康，
簽名檔／署名

各段落超實用句型

說明：畫底線部分的單字可按照個人情況自行替換

開頭 Introduction

❶ **We have received your quotation. Thank you.**

　我們已經收到您的報價單。謝謝您。

❷ **We appreciate your quotation.**

　感謝您的報價。

❸ **Thank you for providing the quotation we requested.**

　感謝您提供我們所要求的報價單。

❹ **I would like to thank you for quoting for <u>the painting work</u> for our <u>new office building</u>.**

　我想要感謝您為本公司新辦公大樓粉刷工程所做的估價。

❺ **In answer to your quotation letter dated on <u>February 12th, 2017</u>, ...**

　為回覆您在 2017 年 2 月 12 日所提供的估價單，……

⑥ We have received your quotation and had a discussion over it <u>this morning</u>.

我們已經收到您的報價了，<u>今天上午</u>進行了討論。

本體 Body ....

❶ It is with great regret that we have to reject your offer.

很抱歉我們必須拒絕您的報價。

❷ Regrettably, we must decline your offer.

很遺憾我們目前必須謝絕您的報價。

❸ I regret not being able to accept your quote.

很遺憾沒有辦法接受您的報價。

❹ Your quoted price is a lot higher than those we received from other suppliers.

貴公司的報價比其他廠商給我們的報價還要高出許多。

❺ We must decline your quotation because we received a better offer.

我們必須拒絕您的報價，因為我們拿到了更優惠的報價。

❻ Unfortunately, we are having second thoughts about the <u>renovation project</u> due to budget reasons, and cannot accept your services at the moment.

很遺憾地，由於預算的關係，我們在考慮取消進行<u>裝潢工程</u>，因此現在無法接受您的服務。

結尾 Closing ....

❶ We will keep in touch with you for future needs.

將來若有需要，我們會與您聯絡。

❷ We look forward to the possibility of doing business with you in the future.

我們期待未來有與貴公司合作的可能性。

❸ We hope to have opportunities to work with you on future projects.

希望有機會能與貴公司共事。

❹ Please feel free to keep in touch for future opportunities.

為了日後合作的可能性，請互相保持聯絡。

❺ We hope you allow us to rely on your services for our <u>other renovation jobs</u> in the future.

未來我們<u>其他的裝修工程</u>，還希望能仰賴您的服務。

❻ If the opportunity arises, we hope that we may still rely on your services.

若有機會，我們希望依舊能仰賴您提供的服務。

 ## 不可不知的實用E-mail字彙

- **decline the quotation** 婉拒報價
- **reject offer** 拒絕報價
- **be unable to accept the offer** 無法接受報價
- **a better price** 更便宜的價格
- **future projects** 未來的合作案
- **future opportunities** 未來合作機會

An Order Change/ Cancellation Request Letter

寫作要點 Key Points:

Step 1: 說明來信目的為變更訂單（提供訂單編號）
Step 2: 說明變更／取消原因，或提供變更內容。
Step 3: 為變更／取消訂單內容致歉。

 實際E-mail範例

寫信 | ▼　　刪除　　回覆 | ▼　　寄件者： Ginny

Dear Sir/Madam,

Introduction

With reference to the order we placed for two hundred suitcases to be delivered on May 15th, we are sorry to inform you that we have to cancel part of our order and you will have to deliver only one hundred suitcases instead of the quantity that we ordered.

Body

This change is resulted from an unexpected clerical error on our part. Considering our long term association with your company, we hope you could waive the fees incurred.

Closing

We sincerely apologize for all the inconvenience that this cancellation may cause.

Thank you for your understanding.

Best regards,
Signature of Sender
Sender's Name Printed

E-mail中譯

寫信▼　刪除　回覆▼　寄件者：Ginny

您好：

開頭

關於我們那份要 5 月 15 日交貨的兩百件行李箱訂單，我們很抱歉通知您，我們必須取消部分訂單，僅交貨一百件行李箱，而非我們原先訂購的數量。

本體

這次會取消訂單是我方意料之外的作業疏失所造成。考慮到我們與貴公司長久以來的合作關係，希望貴公司能不要收取變更訂單的手續費。

結尾

我們對所有這次取消訂單可能造成的不便，致上誠摯的歉意。

感謝您的體諒。

敬祝安康，
簽名檔／署名

各段落超實用句型

說明：畫底線部分的單字可按照個人情況自行替換

開頭 Introduction

❶ **I am writing this letter to request a change in our order.**

我寫這封信是為了請求變更訂單。

❷ **This is to notify you that I need to make changes in the order I confirmed this morning.**

這封信是要通知您，我必須修改今天上午確認過的訂單。

❸ **The purpose of this letter is to inform you that we need to cancel the order we placed on August 17th, 2017.**

這封信的目的是要通知您我們必須取消 2017 年 8 月 17 日的訂單。

❹ This letter is to inform you that our order #70498 needs to be cancelled.

謹以此信通知您，我們編號 70498 的訂單必須取消。

❺ We are sorry to inform you that we have to request the cancellation of our order for 100 printers on September 12th, 2017.

很抱歉通知您，我們必須請求取消 2017 年 9 月 12 日訂購 100 台印表機的訂單。

本體 Body

❶ We would like to cancel the following items from our order.

我們想從訂單中刪除以下品項。

❷ We would like to change the delivery address.

我們希望能更改寄送地址。

❸ We would like to change the quantity of our purchase order as below.

我們希望能更改我們的訂單數量如下。

❹ Due to an error on our part, we need to cancel our purchase order that we placed yesterday.

由於我方疏失，我們必須取消昨天下單的訂購單。

結尾 Closing

❶ I sincerely apologize for any inconvenience that the cancellation may have caused.

我為取消訂單一事所可能造成的不便，致上誠摯的歉意。

❷ Please accept our apology for the inconvenience.

造成不便很抱歉。

❸ We are sorry for all the confusion.

抱歉造成困擾。

❹ Please contact me immediately should you have any questions regarding the change.

若您對此變更有任何疑問，請立刻與我聯絡。

❺ **If you are unable to accept the change, please contact us as soon as possible.**
如果您無法接受變更，請立刻與我們聯絡。

❻ **I look forward to a favorable reply.**
期待能得到您善意的回覆。

 # 不可不知的實用E-mail字彙

- **purchase order** 採購訂單
- **sales order** 業務訂單
- **request change in order** 請求變更訂單
- **request cancellation of order** 請求取消訂單
- **amend the order** 修改訂單
- **alter the order** 修改訂單
- **confirm the order** 確認訂單
- **accept the change** 接受變更
- **change the delivery address** 變更寄送地址
- **change the delivery date** 變更交貨日期
- **change the quantity of order** 變更訂單數量
- **cancel part of the order** 取消部分訂單
- **add items to an order** 在訂單中加入商品
- **delete items from an order** 從訂單中刪除商品
- **an error on our part** 我方疏失
- **clerical error** 作業疏失
- **waive the charges** 免除費用
- **change fees** 變更手續費
- **cancellation fees** 取消手續費

Part6

詢問與請求篇

01 詢問公司資訊

An Information Request Letter

 實際E-mail範例

寫信 ▼	刪除	回覆 ▼	寄件者：	Ginny

Dear Sir/Madam,

Introduction

We are a medium-sized manufacturer of notebook computers looking for computer component suppliers for our production.

Body

We would greatly appreciate your assistance in providing us any information regarding your products and your services.

Closing

We look forward to hearing from you.

Best regards,
Signature of Sender
Sender's Name Printed

E-mail中譯　　寫信▼　刪除　回覆▼　寄件者：Ginny

您好：

....

開頭

我們是一家中等規模的筆記型電腦製造商，目前正在尋找電腦零件供應商。

本體

若您能協助提供我們任何有關貴公司產品或服務的資訊，我們將不勝感激。

結尾

期待得到您的回覆。

....

敬祝安康，
簽名檔／署名

各段落超實用句型

說明：畫底線部分的單字可按照個人情況自行替換

開頭 Introduction

❶ We are a construction company specializing in building large scale public areas.

我們是一間專門負責大型公共區域的營造公司。

❷ We are a clothing manufacturer based in northern Taiwan.

我們是一家位於北台灣的服裝製造公司。

❸ We are an automobile manufacturer looking for auto part suppliers.

我們是一間汽車製造廠，正在尋找汽車零件供應商。

❹ Currently we are seeking additional suppliers for high quality textile materials.

目前我們正在尋求可以提供高品質布料的供應商。

❺ We are currently developing <u>new suppliers</u> for our <u>mass production needs</u>.

我們目前正在開發可以應付本公司大量生產需求的新供應商。

本體 Body

❶ Could you provide us any information with reference to your products and services?

能否請您提供我們任何有關貴公司產品及服務的資訊呢？

❷ We appreciate your assistance in sending us your product catalogues.

感謝您協助提供我們產品型錄。

❸ We would be grateful to receive any of your product catalogue, and brochures.

我們將感激能收到任何貴公司的產品型錄及手冊。

❹ I would be very thankful if you could send us a detailed catalogue regarding your products and services.

若您能寄一份有關貴公司產品及服務的詳細目錄給我們，我將會非常感激。

❺ We would like to receive any information you may have to help us learn more about your services.

我們希望能收到任何可以讓我們對貴公司服務有更多了解的資訊。

❻ We would like a catalogue of your products, preferably with pictures. Any format is fine with us.

我們希望能獲得您的產品目錄，有圖片的最好。任何形式的目錄都歡迎。

結尾 Closing

❶ Please send us your latest product catalogue at the address listed above.

麻煩您將貴公司最新的產品目錄寄到上述的地址給我們。

❷ Thank you in advance for sending the catalogue to us.

先在這裡謝謝您寄目錄給我們。

❸ I look forward to receiving your reply at you earliest convenience.

期待能盡快得到您的回覆。

❹ **We look forward to possibly building a cooperative relationship with your company in the near future.**

我們期待不久的將來能夠與貴公司建立合作關係。

❺ **Should you have any questions regarding sending us your catalogues, please feel free to contact me.**

若有任何關於寄送型錄方面的問題，請儘管與我聯絡。

❻ **We look forward to receiving your product details and can't wait to learn more about what you have to offer.**

我們很期待獲得您的產品細節，也迫不及待想瞭解您所提供的產品。

 不可不知的實用E-mail字彙

- **product brochure** 產品手冊
- **product catalogue** 產品型錄
- **service catalogue** 服務型錄
- **price list** 價目表
- **computer components supplier** 電腦零件供應商
- **fabric material supplier** 布料供應商
- **auto parts supplier** 汽車零件供應商
- **semiconductor components supplier** 半導體零件供應商
- **develop new suppliers** 開發新供應商

02 詢問訂單進度

An Order Status Inquiry Letter

寫作要點 Key Points:

Step 1: 說明來信目的為關心訂單進度（提供訂單編號等資訊）。
Step 2: 請對方告知目前訂單進度。
Step 3: 提醒對方必須如期交貨。

實際E-mail範例

寫信 ▼　　刪除　　回覆 ▼　　　寄件者： Ginny

Dear Sir/Madam,

Introduction

I am writing concerning our order No. 12345 for 3,000 laptop computers scheduled for delivery on May 30th, 2017.

Body

Since we haven't been informed of the exact delivery date of our order, we are very concerned whether you can delivery our order on schedule.

Closing

We hope you understand it is very important for us to have the order delivered on time.

Should you have any difficulty delivering our order by the deadline, please contact us as soon as possible.

Thank you very much.

Best regards,
Signature of Sender
Sender's Name Printed

E-mail中譯　寫信▼　刪除　回覆▼　寄件者：Ginny

您好：

開頭

我是寫信來關心我們訂單編號 12345，預定在 2017 年 5 月 30 日交貨的 3,000 台筆記型電腦的進度。

本體

由於一直沒有收到有關這筆訂單確切出貨日期的通知，因此我們非常擔心這筆訂單是否能如期出貨。

結尾

我們希望您明白，這筆訂單能否準時交貨對我們而言是非常重要的一件事。

萬一您有任何困難，因而無法在截止日前交貨，請立刻與我們聯繫。

非常感謝您。

敬祝安康，
簽名檔／署名

各段落超實用句型

説明：畫底線部分的單字可按照個人情況自行替換

開頭 Introduction

❶ **I am writing this letter to enquire about our order for 100 phone cases.**

我是寫信來詢問有關我們 100 個手機殼的訂單進度。

❷ **This letter is to enquire for our order No. 12345.**

這封信是寫來詢問我們訂單編號 12345 的進度。

❸ **I would like to inquire about the status of my order placed on May 20th, 2017.**

我想詢問我在 2017 年 5 月 20 日這天下單的訂單進度。

❹ With reference to the order we placed for fifty multifunctional dishwashers on July 25th, 2017, ...

有關我們在 2017 年 7 月 25 日訂購五十台多功能洗碗機的訂單，⋯⋯

❺ Regarding the delivery date of my order for 3,000 iPhone chargers, ...

有關我訂購的 3000 個 iPhone 手機充電器，⋯⋯

本體 Body

❶ I haven't been informed of the delivery date.

我還沒收到交貨日期的通知。

❷ I am concerned about the exact date of delivery.

我很關心交貨的確切日期。

❸ We would like to know whether you could make the delivery of our order on schedule.

我們想知道您是否能如期幫我們的訂單出貨。

❹ We would like to confirm with you that our order would be shipped without delay.

我們想要跟您確認，我們的訂單會如期出貨。

❺ Please make sure you can delivery our order on schedule, as we cannot tolerate any delay.

請您務必準時幫我們的訂單出貨，因為我們無法通融任何耽擱。

❻ I do not remember receiving confirmation of the delivery date, and would like to make sure with you that our order will be shipped on schedule.

我不記得有收到送貨日期的確認，所以想與您確認我們訂的貨物能夠準時送貨。

結尾 Closing

❶ I hope you understand that we would not be able to extend the deadline.

希望您理解我們無法延長交貨日期。

❷ We hope you won't let us down in terms of the delivery date.

希望您在交貨日期上不會讓我們失望。

❸ Please let us know immediately if you have difficulty in delivering our order on time.

如果您無法準時出貨，請立刻通知我們。

❹ I hope you would not postpone the delivery of our order.

我希望您不會延遲我們的訂單交貨。

❺ Should you have any problems regarding the delivery date, please contact us as soon as possible.

萬一您有任何關於出貨日期的問題，請立刻與我們聯繫。

❻ Please understand that we cannot tolerate delays in the delivery and would highly appreciate your cooperation.

請瞭解，我們無法接受送貨過程延遲，若您能配合，我們將感激不盡。

 不可不知的實用E-mail字彙

- **the status of the order** 訂單進度
- **postpone the delivery date** 延遲交貨日
- **make the delivery on schedule** 如期交貨
- **extend the deadline** 延長交貨期限
- **inquire about the status of order** 詢問訂單進度
- **check the status of order** 查詢訂單進度

03 詢問款項進度

A Status of Payment Inquiry Letter

寫作要點 Key Points:

Step 1: 感謝對方的訂單，並提供訂單編號。
Step 2: 說明付款相關事宜。
Step 3: 提醒對方如期付款。

實際E-mail範例

| 寫信 ▼ | 刪除 | 回覆 ▼ | 寄件者： Ginny |

Dear Sir/Madam,

Introduction

Thank you for ordering computer components from our company. Your order No. 12345 has been accepted and processed for delivery.

Body

Please kindly be advised that the payment of your order is expected to be made in full settlement within 3 days on receipt of this letter, otherwise we will be unable to deliver your order.

Closing

I would greatly appreciate it if you could notify us once you settle the payment, and keep the remittance slip for future reference.

Thank you for your business.

Best regards,
Signature of Sender
Sender's Name Printed

E-mail中譯

寫信│▼　刪除　回覆│▼　寄件者：Ginny

您好：

開頭

感謝您向本公司訂購電腦零件。我們已經受理您編號為 **12345** 的訂單，並已進入交付運寄的程序。

本體

在此通知您，您的帳款應在收到此信後三天內全額結清，否則我們將無法為您出貨。

結尾

若您能在結清款項後通知我們，我們將不勝感激。並請將匯款單留存，以備日後查驗。

謝謝您的惠顧。

敬祝安康，
簽名檔／署名

各段落超實用句型

說明：畫底線部分的單字可按照個人情況自行替換

開頭 Introduction

❶ **Thank you for ordering our products.**
感謝您訂購本公司產品。

❷ **Your order for 1,000 travel plug adapters has been accepted.**
您訂購一千個旅行用轉接頭的訂單已經成立了。

❸ **Your order is now ready for delivery.**
您的訂單現在已經進入出貨程序。

❹ **We are pleased to inform you that your order has been processed and sent for delivery.**
我們很高興通知您，我們已經受理您的訂單，並已交付寄運。

❺ I am writing this letter regarding your order No. 12345, for 1,000 tons of rubber material, ...

此信乃有關您訂購的 1000 噸橡膠原料，編號 12345 的訂單，……

❻ This message is to let you know that your order, No. 12416, is all set and ready to ship.

這封訊息是要告知您，您的訂單 12416 號已經準備完成，隨時可發貨。

本體 Body

❶ I would like to confirm with you that the payment of your order is due on January 12th, next Wednesday.

我想跟您確認，您的訂單帳款應付日為 1 月 12 日，也就是下週三，。

❷ Please could you complete the payment by the end of this week?

可否麻煩您在本週內結清帳款？

❸ Please allow me to remind you that you are expected to make the payment within 7 days after receipt of your commodities.

請容許我在此提醒您，您必須在貨到七天之內付清帳款。

❹ Please send the full payment for your purchase within three days on receipt of this letter.

請您在收到這封信後三天內支付購買商品的全額款項。

❺ Please be notified that we would not deliver your order until the payment is settled.

謹在此通知您，款項結清之前，我們無法將您的訂單交付運寄。

❻ We will, unfortunately, be unable to send the ordered items until we receive the payment in full.

很遺憾地，沒有收到全額付款之前，我們無法送出您訂的貨物。

結尾 Closing

❶ Please keep me notified when the remittance is made.

匯款完成後請通知我。

❷ Please send me a copy of the remittance slip for record.

請您將匯款單影本寄給我，以茲記錄。

❸ Please notify us when you settle the payment.
請在付款完成後通知我們。

❹ According to our contract, you are expected to make the payment in full settlement of your purchase within a month.
根據我們的合約，您應該要在訂貨一個月之內匯款結清全額款項。

❺ Attached herewith is the invoice of your order.
隨函附上您的訂單收據。

❻ We will ship the order as soon as we receive confirmation of your payment.
一旦確認您已付款，我們將立即送出貨物。

 # 不可不知的實用E-mail字彙

- **make the payment** 付款
- **complete the payment** 結清帳款
- **remittance slip** 匯款單
- **in full settlement** 全額結清
- **late payment** 延遲付款
- **interest charge** 利息費用
- **for future reference** 以備日後查驗

157

04 請求對方介紹產品／服務

A Product Information Request Letter

寫作要點 Key Points:

Step 1: 表示對於對方產品或服務有興趣。
Step 2: 請對方提供相關產品或服務之資訊。
Step 3: 提供聯絡方式，並期待對方回覆。

 實際E-mail範例

| 寫信 ▼ | 刪除 | 回覆 ▼ | 寄件者： Ginny |

Dear Sir/Madam,

Introduction

We are a clothing manufacturer based in Taipei, seeking for a textile material supplier. We have heard that your company specializes in producing various kinds of cloth of high quality; therefore we are interested in your products.

Body

We would highly appreciate it if you could kindly send us a brochure or a catalogue so that we can have a clearer idea about your products.

Closing

Below is the contact information of our company.

Taipei Fashion Ltd.
Office number: (02)2123-4567
Office address: 7F, 203, Sec. 2, Chungshan N. Rd., Taipei

Please send us any related information regarding your products directly to the address shown above.

Thank you in advance for your time in this matter.

Best regards,
Signature of Sender
Sender's Name Printed

E-mail中譯　　寫信｜▼　　刪除　　回覆｜▼　　寄件者：Ginny

您好：

開頭

我們是一家位在臺北的服裝製造廠，目前正在尋求布料供應商。我們耳聞貴公司專門生產各種高品質的布料，因此我們對貴公司的產品很有興趣。

本體

如果您能寄給我們一份產品手冊或型錄，讓我們能更清楚地了解貴公司的產品，我們將不勝感激。

結尾

以下為本公司聯絡資訊：

臺北時尚股份有限公司

公司電話：(02)2123-4567

公司地址：台北市中山北路二段 203 號 7 樓

請直接將任何有關貴公司產品的資訊寄到以上地址。

在此先向您致謝。

敬祝安康，
簽名檔／署名

# 各段落超實用句型

說明：畫底線部分的單字可按照個人情況自行替換

開頭 Introduction

❶ I am writing this letter to enquire for your latest range of <u>sanitary ware</u>.

我是寫這封信來詢問有關貴公司最新一系列的衛浴設備。

❷ We are very interested in your new range of <u>kitchenware advertised on television</u>.

我們對貴公司在電視上廣告的新系列廚房用具很有興趣。

❸ We are very interested in buying <u>household appliances</u> from your company.

我們有興趣購買貴公司的家電用品。

❹ As we are planning to replace the <u>central air conditioning system</u> of our <u>office building</u>, we would like to learn more about your products and services.

由於我們正計劃要更換本公司辦公大樓的中央空調系統，我們希望能知道更多有關貴公司產品及服務的資訊。

❺ With this letter I would like to inquire about your <u>electronic products</u>.

我想經由這封信詢問有關貴公司電子產品的資訊。

本體 Body

❶ I would like to request you to send us detailed information on your <u>consumer electronics</u>.

我想請您寄給我們貴公司消費類電子產品的詳細資訊。

❷ We would like to have more information about your new launched products.

我們想要取得更多關於你們新上市的產品資訊。

❸ We would like to invite your sales representative to our company and introduce your products.

我們想邀請貴公司的業務代表到我們公司來介紹您的產品。

❹ We would like to request a meeting with your sales representative for a detailed product introduction.

我們想跟貴公司業務代表碰面，請他為我們做詳細的產品介紹。

❺ Please send us a soft copy of your product catalogue if possible.

如果可以的話，請寄產品目錄的電子檔給我們。

結尾 Closing

❶ We look forward to your reply.

期待您的回覆。

❷ Please call 1234-5678 for Jennifer regarding this matter.

有關此事請致電 1234-5678 找珍妮佛。

❸ Please send the information requested to the following address.

請將我們請求的資料寄到以下地址。

❹ We would greatly appreciate it if we could receive it within this week.

如果可以在本週內收到，我們將感激不盡。

❺ We hope to receive the catalogue by our monthly department meeting held next Monday.

我們希望能在下週一舉行部門月會之前收到型錄。

 不可不知的實用E-mail字彙

- **soft copy** 電子檔
- **hard copy** 印刷本
- **request a catalogue** 索取目錄
- **product introduction** 產品介紹
- **new product presentation** 新品發表
- **new product launch** 新品上市

A Refund Request Letter

Step 1: 簡短說明來信退費之目的。
Step 2: 直接表示請求退費，並表明原因。
Step 3: 請對方盡速處理退費，並給予回覆。

 實際E-mail範例

   寄件者： Ginny

Dear Sir/Madam,

Introduction

We have received our commodities that we ordered from you on July 17th, 2017, and sadly found the quality of the fabrics far from our expectation.

Body

With great regret, we are writing this letter to inform you that we are returning the commodities for a full refund.

Closing

We will send the commodities back to you within two days and hope you will reimburse us for the delivery cost.

Best regards,
Signature of Sender
Sender's Name Printed

E-mail中譯　寫信｜▼　刪除　回覆｜▼　寄件者：Ginny

您好：

開頭

我們已經收到 2017 年 7 月 17 日向貴公司訂購的商品，並且難過地發現布料的品質遠不如我們的期待。

本體

很遺憾地在此寫信通知您，我們將退還商品，並請求全額退費。

結尾

我們會在兩日內將商品寄還，並希望貴公司能補償我們運費。

敬祝安康，
簽名檔／署名

各段落超實用句型

說明：畫底線部分的單字可按照個人情況自行替換

開頭 Introduction

❶ **It is with great regret that we must request for a refund for our order.**
非常遺憾地，我們必須請求訂單退費。

❷ **We would like to return your products.**
我們想要退回您的產品。

❸ **On behalf of JBS Company, I am writing this letter to notify you that we are cancelling our order No. 12345 and request for a refund.**
謹代表 JBS 公司寫信通知您，我們將取消我們編號 12345 的訂單，並請求退費。

❹ **We would like to return our commodities for a full refund.**
我們希望能全額退還商品。

❺ **We would like to issue a full refund.**
我們想要求全額退款。

❻ I am writing this email to express dissatisfaction with my order, <u>No.</u> <u>12412</u>, which I would like to return.

我寫這封信是要表達對我訂的 12412 號訂單不滿，想退貨。

本體 Body

❶ We are not happy with <u>the product</u> and demand a refund.

我們對這商品不滿意，請求退款。

❷ We would like to request for a refund of the <u>duplicate payment</u>.

我們想請貴公司退回重複給付的款項。

❸ <u>The quality</u> is very different from the samples you provided us.

商品品質跟你們提供給我們的樣品相差甚遠。

❹ We regret to inform you that we must return the product because <u>it didn't work at all</u>.

我們很遺憾通知您，我們必須退回商品，因為它完全不能使用。

❺ We need to cancel the order as you <u>kept postponing the delivery date</u>.

我們必須取消訂單，因為貴公司一直拖延交貨日期。

❻ Unfortunately, the items I received did not look anything like their pictures on the website.

很遺憾地，我收到的貨品與網站上圖片的樣子完全不像。

結尾 Closing

❶ Thank you for your prompt attention to this matter.

請您即刻處理此事，謝謝。

❷ Please refund our payment at your earliest convenience.

請您盡快退回我們的帳款。

❸ We look forward to receiving our refund promptly.

我們期待能很快地收到我們的退款。

❹ Please return the payment within the next <u>7 days</u>.

請在七日內退回帳款。

❺ **Please contact our staff should you have any questions regarding this matter.**
若您有任何關於此事的問題請與本公司員工聯絡。

❻ **Please see the attached refund request form and settle the matter as soon as possible.**
請見附件的退款申請單，並盡快處理。

 不可不知的實用E-mail字彙

- **demand a refund**　要求退款
- **request for a refund**　要求退款
- **return for a full refund**　全額退回（商品）
- **issue a full refund**　申請全額退款
- **reimbursement of delivery charges**　運費補償

06 請求提供樣品

A Sample Request Letter

實際E-mail範例

寫信 ▼	刪除	回覆 ▼	寄件者： Ginny

Dear Sir/Madam,

Introduction

We are a notebook computer manufacturer based in the south of Taiwan currently seeking new computer components suppliers. We learn that your company specializes in producing high quality computer components, and are very interested in buying products from you.

Body

With this letter we would like to make a request that you kindly provide your sample products for consideration.

We wish to have enough time to try out the samples before deciding whether we would like to buy them in our department meeting held on April 20th.

Closing

Looking forward to receiving the samples promptly.

Thank you very much.

Best regards,
Signature of Sender
Sender's Name Printed

E-mail中譯

寫信 ▼　刪除　回覆 ▼　寄件者：Ginny

您好：

開頭

我們是一家位於南台灣的筆記型電腦製造公司，目前正在尋找新的電腦零件供應商。我們得知貴公司專門生產高品質電腦零件，因此有興趣購買貴公司的產品。

本體

透過此信我們想要提出一個請求，就是希望貴公司能提供樣品，讓我們做採購考量。

我們將會在四月 20 日舉行的部門會議中決定是否採購，因此在那之前我們希望能有足夠時間試用樣品。

結尾

期待能很快收到樣品。

非常感謝您。

敬祝安康，
簽名檔／署名

各段落超實用句型

說明：畫底線部分的單字可按照個人情況自行替換

開頭 Introduction

❶ **We are very interested in the** <u>new coffee machine</u> **you just launched.**
我們對貴公司剛推出的<u>新咖啡機</u>很有興趣。

❷ **We would like to try out your new** <u>auto shredder</u>**.**
我們想試用貴公司的新<u>電動碎紙機</u>。

❸ **We are very interested in trying out the new functions of your** <u>photocopier</u>**.**
我們非常有興趣試試貴公司<u>影印機</u>的新功能。

❹ I have received your new product introduction letter and would like to try out this product in person.

我已經收到貴公司的新品介紹信，並且很想親自試用這個產品。

❺ We are eager to try out your new launched product.

我們非常希望能試用您新推出上市的產品。

本體 Body

❶ We would like to receive some samples to try out.

我們希望能收到一些樣品試用。

❷ We would like to request for some free samples.

我們想要請你們提供一些免費的試用品。

❸ We would greatly appreciate it if you could kindly provide us a sample product to try out.

如果您能提供我們一些樣品試用，我們會非常感謝。

❹ Could you please send us sample copies for our preview?

能否請您寄一些試閱本讓我們預讀？

❺ We need to try out your sample products in order to decide what items to purchase.

我們需要試用您的樣品，以便決定該購買哪些品項。

❻ We would be delighted if you could send us some sample products, as they would help us greatly in making the final decision.

若您能寄送一些樣品給我們，我們將感激不盡。樣品對我們做出最終決定有很大的幫助。

結尾 Closing

❶ We hope to receive the samples within this week, if possible.

如果可以的話，我們希望可以在本週內收到樣品。

❷ Please send the samples at your earliest convenience.

麻煩您盡快寄出樣品。

❸ Please send the samples directly to the following address.

請直接將樣品寄至以下地址。

❹ Please call <u>1234-5678</u> for <u>Simon</u> for any questions you may have.

有任何問題請播打 <u>1234-5678</u> 找<u>賽門</u>。

❺ Should you have problems in providing sample products, please feel free to let us know.

若您無法提供樣品，請讓我們知道。

❻ Attached is a detailed list of the products for which we would like a sample. Please feel free to let me know if you cannot provide samples for any of them.

附上的產品清單是我們希望收到的樣品內容。若無法提供樣品，請不吝告知我。

 不可不知的實用E-mail字彙

• **sample product** 樣品，試用品
• **sample copy** 試閱本
• **free sample** 免費試用品
• **try out** 試用
• **try out the new functions** 試用新功能

07 請求提供建議

An Advice Request Letter

寫作要點 Key Points:

Step 1: 簡短具體說明問題。
Step 2: 請對方提供建議。
Step 3: 請求對方回覆。

 實際E-mail範例

| 寫信 ▼ | 刪除 | 回覆 ▼ | 寄件者： Ginny |

Dear Sir/Madam,

Introduction

As you may already know, we are thinking about opening a branch restaurant in Shanghai and we are carefully evaluating our business expansion plan.

Body

This is going to be our first overseas branch. Most important of all, we are not familiar with the related policies and regulations in China. Therefore, I would appreciate it if you could kindly give me some advice about expanding my restaurant business to China.

Closing

I am looking forward to your kind reply.

Thank you in advance for your assistance in this matter.

Best regards,
Signature of Sender
Sender's Name Printed

E-mail中譯

寫信▼　刪除　回覆▼　寄件者：Ginny

您好：

開頭

如您所知，我們正在考慮要在上海開設一家餐廳分店，而且我們正很審慎地評估我們的事業拓展計劃。

本體

這將會是我們第一間海外分店。最重要的是，我們對中國的相關政策與規定並不熟悉。因此如果您能針對我們拓展餐飲事業至中國大陸一事，給予一些建議，我們將不勝感激。

結尾

期待您的回覆。

在此先感謝您在這件事上給予的協助。

敬祝安康，
簽名檔／署名

各段落超實用句型

說明：畫底線部分的單字可按照個人情況自行替換

開頭 Introduction

❶ We are currently thinking about expanding our business to China.
我們目前正計劃將業務拓展至中國。

❷ We are planning to acquire BCB Company for business expansion.
我們目前正計劃併購 BCB 公司，以擴大公司經營。

❸ We are carefully evaluating the possibility of merging with BCB Company to grow our business.
我們目前正在評估合併 BCB 公司以擴大公司經營的可能性。

❹ We are confronting a contract dispute with another company.
我們目前正面臨與另一間公司的合約糾紛。

171

⑤ We intend to open a new branch restaurant in <u>southern Taiwan</u>.

我們有意在<u>南台灣</u>開設一間餐廳分店。

⑥ I am writing to seek advice in a legal matter regarding <u>company expansion affairs</u>.

我寫這封信，是希望獲得關於<u>公司擴張</u>的法律方面的建議。

本體 Body

❶ I would like to ask you for some <u>legal</u> advice.

我想向您徵詢一些<u>法律方面</u>的建議。

❷ We are in desperate need of your advice on our <u>business expansion</u> plan.

我們極度需要您給予我們有關<u>業務擴展</u>計劃的建議。

❸ We would like to have a meeting with you to discuss this plan.

我們希望能跟您開個會，討論一下這個計劃。

❹ We would greatly appreciate it if you could provide us some advice on the new project.

若您能提供我們一些關於這項新企劃的建議，我們將不勝感激。

❺ We would be grateful if you could share opinions on this matter with us.

如果您能與我們分享您在這個問題上的看法，我們將非常感激。

❻ As none in our company are knowledgeable in this regard, your guidance will be greatly appreciated.

由於我們公司沒有人對此方面瞭解，我們將非常感激您的建議。

結尾 Closing

❶ Thank you for your assistance.

感謝您的協助。

❷ Look forward to hearing from you.

期待您的回覆。

❸ We hope to have a personal meeting with you.

我們希望能與您私下開個會。

❹ I have attached some documents regarding this project for your reference.

我已附上一些有關此專案的文件讓您參考。

❺ I will call your office today to see when you might be available for a brief conversation.

我今天會打電話到您公司，看看您何時有空可以與我們談談。

❻ If you are unavailable to meet up with us, we would appreciate any recommendations on qualified professionals who may aid us in this matter.

若您無法與我們見面，但可以推薦我們一些能夠幫助我們的專業人士，我們將感激不盡。

不可不知的實用E-mail字彙

- **ask for advice** 請求建議
- **provide advice** 提供建議
- **offer suggestions** 提供建議
- **offer recommendations** 提供建議
- **share one's opinions in sth.** 分享對於某事的看法

A Shipment Delay Request Letter

寫作要點 Key Points:

Step 1: 明白告知對方無法如期交貨。
Step 2: 簡短說明無法如期交貨的原因。
Step 3: 致歉，請求對方同意延後交貨。

 實際E-mail範例

| 寫信 ▼ | 刪除 | 回覆 ▼ | | 寄件者： | Ginny |

Dear Sir/Madam,

Introduction

With reference to your order, No. 12345, for 3,000 electric toothbrushes, scheduled to be delivered on January 12th, we regret to have to inform you that we might not be able to meet the delivery date.

Body

The reason of the entire delay was caused by an unexpected problem occurred in our delivery process.

Closing

Please accept our apology for this delay.

We will manage to deliver your order as soon as possible.

Best regards,
Signature of Sender
Sender's Name Printed

E-mail中譯

寫信│▼　　刪除　　回覆│▼　　寄件者：Ginny

您好：

開頭

有關您編號 12345，訂購三千支電動牙刷，預定於 1 月 12 日交貨的訂單，我們很遺憾必須通知您，我們可能無法如期交貨。

本體

延遲交貨的原因乃是因為我們出貨流程突然出了點問題。

結尾

請接受我們為延遲交貨向您致上的歉意。

我們會設法盡快寄出您的訂單。

敬祝安康，
簽名檔／署名

各段落超實用句型

說明：畫底線部分的單字可按照個人情況自行替換

開頭 Introduction

❶ **I am writing this letter to inform you that we will have to delay your delivery.**

這封信的目的是要通知您，我們將必須延遲交貨。

❷ **We regret to inform you that we are unable to deliver your order on schedule.**

很抱歉通知您，我們恐怕無法如期交貨。

❸ **To our regret, the delivery of your order will be delayed for a couple of days.**

很抱歉，您的訂單將會延遲幾天出貨。

❹ With reference to your order No. 12345, we are notifying you that some items are not available for shipping immediately.

有關您編號 12345 的訂單，我們必須通知您有些品項目前無法立刻交貨。

❺ With great regret, this letter serves to inform you of the delay of your order.

很遺憾地以這封信通知您，您的訂單將延遲交貨。

本體 Body

❶ We mixed up your delivery dates with others.

我們把您的交貨日期跟其他的搞混了。

❷ We had some problems with our delivery system.

我們的出貨系統出了一點問題。

❸ The delay is caused by inventory shortage.

存貨短缺造成交貨延遲。

❹ Our supplier failed to provide us sufficient components for our production.

我們的供應商無法提供我們足夠的零件讓我們生產。

❺ We are currently very shorthanded to deal with an unusually large number of orders at present.

我們目前人力非常不足，無法處理比平常更大量的訂單。

❻ Our normal delivery schedule is seriously disrupted by an unusually large number of orders during the holiday season.

我們正常的出貨流程被假期帶來的大量訂單給打亂了。

結尾 Closing

❶ Unfortunately, we will have to postpone the delivery until February 22nd.

很遺憾我們將必須延遲交貨時間至 2 月 22 日。

❷ We're sorry to notify you that your shipment will be delayed for two weeks.

很抱歉通知您，您的出貨將會延遲兩星期。

❸ **Please accept our sincere apology for the entire delay.**
請接受我們對延遲交貨的誠摯歉意。

❹ **We will expedite delivery to you once we are in the position to do so.**
一旦情況允許我們就會立刻出貨給您。

 不可不知的實用E-mail字彙

- **delay delivery** 延遲交貨
- **delay shipment** 延遲交貨（運送）
- **make delivery** 交貨
- **lack of inventory** 缺貨
- **expedite delivery** 發貨
- **delivery process** 出貨流程
- **delivery system** 出貨系統
- **unusually large number of orders** 異於平常的大量訂單
- **normal delivery schedule** 正常出貨時間

09 請求延後付款

A Payment Delay Request Letter

實際E-mail範例

寫信 ▼	刪除	回覆 ▼	寄件者： Ginny

Dear Sir/Madam,

Introduction

I am writing this letter with regard to the payment due on this Friday. It is with great regret that I have to inform you that the payment will be delayed for a couple of days.

Body

Please understand that this delay is not intended. We have just moved to our new office and the payment bills and invoice documents were all mixed up due to the office relocation.

We shall settle the payment by October 19th, 2017. Please do contact us immediately if you do not receive the payment at the appointed time.

Closing

We are very sorry for the inconvenience due to this payment delay, and we assure you that we will not allow this kind of mistake to happen again.

Thank you for your understanding.

Best regards,
Signature of Sender
Sender's Name Printed

E-mail中譯　寫信|▼　刪除　回覆|▼　寄件者：Ginny

您好：

開頭

我寫這封信是有關這個星期五到期的帳款。很遺憾必須通知您，這筆帳款將會延遲數天支付。

本體

請明白我們並非蓄意延遲支付貨款。我們辦公室最近搬遷，因此所有的帳單和收據文件都因為辦公室遷移而搞混。

我們將會在 2017 年 10 月 19 日之前結清帳款。屆時若您未收到帳款，請務必立刻跟我們聯絡。

結尾

我們對這次延遲付款所造成的不便感到萬分抱歉，而且我們向您保證，不會再讓相同的錯誤再次發生。

感謝您的體諒。

敬祝安康，
簽名檔／署名

 各段落超實用句型

説明：畫底線部分的單字可按照個人情況自行替換

開頭 Introduction

❶ **With this letter we have to inform you that we will have to delay our payment.**

很抱歉寫這封信通知您，我們將會延遲支付貨款。

❷ **I regret to inform you that our payment will be delayed.**

很抱歉要通知您，我們的帳款將會延遲支付。

❸ **We understand that our payment is due on September 15th.**

我知道我們的帳款在將於 9 月 15 日到期。

❹ We are well aware that we need to settle our payment within 30 days after receipt of our order.

我們很清楚知道我們必須在到貨三十天內結清帳款。

❺ We are sorry for not being able to make the payment on the expected date.

我們很抱歉不能如期支付帳款。

本體 Body

❶ We would like to request an extension to complete our payment.

我們想要請求延期支付帳款。

❷ We need an extension on our payment.

我們需要延長我們的繳款期限。

❸ We are encountering some cash flow problems.

我們目前正面臨一些現金流問題。

❹ The payment delay is due to the office relocation.

延遲付款是因為辦公室遷移所致。

❺ Due to staff change, we're unable to process all the invoices and payments in time.

由於人事變動的關係，我們無法及時處理所有的發票和帳款。

❻ Unfortunately, we are unable to process the payment as scheduled.

很遺憾地，我們無法如期處理付款。

結尾 Closing

❶ We will manage to settle the payment by the end of the month.

我們會設法在本月底之前結清帳款。

❷ We ensure that you will receive the payment by April 30th.

我們保證您會在 4 月 30 日之前收到帳款。

❸ We will remit the outstanding payment to your account within the next ten days.

我們將會在未來十天之內將未結清款項匯到您的帳戶。

❹ Please kindly be advised that the payment will be settled by <u>5 pm tomorrow</u>.

請留意帳款會在<u>明天下午 5 點</u>前付清。

❺ Our financial department will complete the payment in <u>two days</u>.

我們的財務部門會在<u>兩天</u>內結清帳款。

❻ We will complete the payment during <u>the next week</u>. Please feel free to let us know if you still have not received the payment then.

我們<u>下週</u>將會完成付款。若到時候您還沒有收到付款，請不吝告知我們。

 不可不知的實用E-mail字彙

- **extension on payment** 延長繳款期限
- **grant a deferment** 同意延期
- **request an extension** 請求延期
- **cash flow problem** 資金周轉問題
- **extend the due date** 延長繳款截止期限

Part7

各式通知篇

A Shipping Notice Letter

寫作要點 Key Points:

Step 1: 感謝對方下訂單，告知對方訂單已經出貨。
Step 2: 提醒對方到貨時間。
Step 3: 請對方於收貨後告知，並感謝對方合作。

 實際E-mail範例

| 寫信 ▼ | 刪除 | 回覆 ▼ | 寄件者： Ginny |

Dear Sir/Madam,

Introduction

Thank you for ordering products from our company. We are pleased to inform you that your order for 500 hairdryers has been sent to you.

Body

Please be kindly advised that your commodities will arrive at your appointed address in two days.

Closing

Please let us know immediately if you don't receive your order at the expected time.
We thank you for your business and look forward to doing business with you again.

Best regards,
Signature of Sender
Sender's Name Printed

E-mail中譯　寫信 ▼　刪除　回覆 ▼　寄件者：Ginny

您好：

開頭

感謝您向本公司訂購產品。我們很高興通知您，您訂購的五百支吹風機已經寄送給您。

本體

請注意您的商品將會在兩天內送達您指定的地址。

結尾

如果您在預定時間未收到您訂購的貨品，請立刻通知我們。

感謝您的惠顧，並期待能再次為您服務。

敬祝安康，

簽名檔／署名

各段落超實用句型

說明：畫底線部分的單字可按照個人情況自行替換

開頭 Introduction

❶ **Thank you for ordering products from our company.**

感謝您向本公司訂購產品。

❷ **Thank you for placing an order with us.**

感謝您向我們下訂單。

❸ **Thank you for your order.**

感謝您的訂單。

❹ **Thank you for your purchase order dated September 27th.**

感謝您 9 月 27 日的訂單。

❺ **Thank you for the order of computer components you have placed with us.**

感謝您向我們訂購電腦零件的訂單。

本體 Body

❶ Your commodities have been shipped to be received no later than April 29th.

您的商品已經出貨,並會在 4 月 29 日前送達。

❷ Your order will be shipped to you COD within five working days.

您訂的貨會在五個工作天之內,以貨到付款的方式寄送給您。

❸ Your order will arrive at your appointed address in 24 hours.

您的商品將會在 24 小時之內送達您指定的地址。

❹ Your order is now ready for pickup at your appointed convenience store.

您現在已經可以到您指定的超商領取您的商品了。

❺ Should you not receive your commodities at the expected time, please contact us immediately.

如果您沒有在預定時間收到商品,請立刻與我們聯絡。

❻ We are glad to inform you that the items you ordered have been shipped and will arrive in 2-5 working days.

很開心告知您,您所訂的產品已經出貨,將會在 2-5 個工作天內抵達。

結尾 Closing

❶ We thank you for your business.

感謝您的惠顧。

❷ Please let us know if you're interested in any of our other products.

如果您對本公司任何其他產品有興趣,請與我們聯絡。

❸ Please feel free to contact us whenever you need any information regarding our products.

無論何時只要您需要任何有關本公司的產品資訊,都歡迎您與我們聯繫。

❹ We look forward to your continuous business with us in the future.

期待未來您能繼續惠顧。

❺ We look forward to serving you again soon.
我們期待很快能再次為您服務。

❻ Thank you for shopping with us. We hope you will drop by our website soon!
謝謝您的惠顧。希望您不久後能夠再拜訪我們的網站！

 不可不知的實用E-mail字彙

- **appointed address** 指定地址
- **working days** 工作日
- **appointed convenience store** 指定超商
- **within 24 hours** 在 24 小時之內
- **COD= cash on delivery** 貨到付款

02 缺貨通知

An Out-of-Stock Notice Letter

寫作要點 Key Points:

Step 1: 感謝對方訂購或諮詢商品。
Step 2: 告知對方所訂購商品缺貨。
Step 3: 對缺貨表示抱歉，並表示貨到時會通知。

 實際E-mail範例

寫信 ▼　　刪除　　回覆 ▼　　　寄件者：　Ginny

Dear Sir/Madam,

Introduction

Thank you for placing an order with us.

Body

Unfortunately, we must inform you with this letter that we are unable to accept your order because the electric toothbrush that you ordered is currently out of stock.

Closing

Please accept our apology for cancelling this order.

We will let you know as soon as the product is back in stock.

Look forward to future opportunities to provide our service to you.

Best regards,
Signature of Sender
Sender's Name Printed

E-mail中譯

寫信｜▼　刪除　回覆｜▼　寄件者：Ginny

您好：

開頭

感謝您向本公司下訂單。

本體

很可惜，我們必須以這封信通知您，我們無法受理您的訂單，因為您所訂購的電動牙刷目前缺貨中。

結尾

我們必須取消這份訂單，請接受我們的道歉。

該產品貨到時我們會立刻通知您。

期待未來有機會為您提供服務。

敬祝安康，
簽名檔／署名

各段落超實用句型

說明：畫底線部分的單字可按照個人情況自行替換

開頭 Introduction

❶ Thank you for your enquiry for our kitchen extractor, KE1233.
感謝您咨詢我們的廚房抽油煙機，型號 KE1233。

❷ Thank you for your interest in our new range of kitchenware.
感謝您對我們新系列的廚具感興趣。

❸ With reference to your enquiry for our new line of lipstick, we regret to tell you that it's currently unavailable.
有關您對本公司新系列唇膏的詢問，我們很遺憾地通知您，該系列目前沒有存貨。

❹ We hereby inform you that the trousers you ordered on December 12th, 2017 are now out of stock.

我們在此通知您，您在 2017 年 12 月 12 日訂購的褲子目前缺貨。

❺ We are unable to deliver your order because it is out of stock.

我們現在無法寄出您的訂單，因為訂單產品目前缺貨。

本體 Body

❶ The size you ordered is now unavailable.

您所訂購的尺寸目前沒有貨。

❷ We were left with extremely low levels on the inventory of the item that you ordered.

您訂購的產品，僅剩下非常少的存貨。

❸ We do not have enough stock to meet your order at present.

我們目前沒有足夠存貨可以符合您的訂單。

❹ Unfortunately the items will not be restocked anytime soon.

很遺憾這些商品短期內不會進貨。

❺ We are making every effort to get the items back in stock.

我們正積極補貨中。

❻ New stocks will be available by the end of the month.

新貨將會在月底前到。

❼ We are sorry to inform you that the specific version you are asking for has been sold out.

很遺憾必須通知您，您想要的版本已經賣光了。

結尾 Closing

❶ We will cancel your order and you will get a full refund within 3-5 working days.

我們會取消您的訂單，而您將會在三～五個工作天內拿到全額退款。

❷ We hope our updated product catalogue contains something that interests you.

希望我們最新的產品型錄有讓您感興趣的產品。

❸ **We appreciate your continued patronage and look forward to serving you.**
我們感謝您繼續惠顧，並期待能為您服務。

❹ **We apologize for not being able to accept your order this time.**
我們為這次無法接受您的訂單向您致歉。

❺ **If any other version of the product interests you, do let us know and we will be glad to serve you.**
若您對此產品的其他版本感興趣，請告知我們，我們很樂意為您服務。

 # 不可不知的實用E-mail字彙

- **out of stock** 缺貨
- **currently not available** 目前缺貨
- **inventory shortage** 存貨短缺
- **replenish the stock** 補貨
- **back in stock** 重新到貨
- **waiting list** 等候名單

03 休假通知

A Leave Notice Letter

寫作要點 Key Points:

Step 1: 發出休假通知，並告知休假時間。
Step 2: 聯絡休假期間相關業務事宜。
Step 3: 感謝對方體諒，並對休假造成的不便致歉。

 實際E-mail範例

寫信｜▼　　刪除　　回覆｜▼　　　寄件者：　Ginny

Dear Sir/Madam,

Introduction

This letter is to inform you that I will take a week off for annual leave and will not be in the office from January 7th to January 14th.

Body

Please note that our Deputy Manager Ms. Lillian Yeh is going to cover my position while I am on leave. Should you have any questions and need any assistance, please contact her directly.

Closing

Thank you for your continued support.

I will get in touch with you as soon as I return to work.

Best regards,
Signature of Sender
Sender's Name Printed

E-mail中譯

寫信│▼　刪除　回覆│▼　寄件者：Ginny

您好：

開頭

謹以此信通知您，我將從 1 月 7 日至 1 月 14 日休年假，並且在這段期間不會進公司。

本體

在此通知您，在我休假期間，我們的副經理葉麗安將會代理我的職務。如果您有任何問題或需要任何協助，請直接與她聯繫。

結尾

感謝您一直以來的支持。

我一回到工作崗位就會立刻與您聯絡。

敬祝安康，
簽名檔／署名

 各段落超實用句型

說明：畫底線部分的單字可按照個人情況自行替換

開頭 Introduction

❶ **I am writing this letter to inform you that we are not open for business from January 12ᵗʰ to January 15ᵗʰ.**

謹以此信通知您，我們在 1 月 12 日至 1 月 15 日這段期間將不會營業。

❷ **Please note that we will be closed for a week, starting from January 12ᵗʰ.**

在此通知您，我們將從 1 月 12 日開始店休一週。

❸ **With this letter I would like to notify you that I will take three days off from January 12ᵗʰ for personal leave.**

謹以此信通知您，我將會從 1 月 12 日起休三天事假。

❹ Please be kindly advised that I will be taking my maternity leave and parental leave altogether for six months, starting from May 1st, 2017.

在此通知您，我將會從 2017 年 5 月 1 日起，連同產假及育嬰假，連續休假六個月。

❺ I am writing to notify you that I will take my marriage leave from June 1st to June 12th.

我特地寫信通知您，我將在 6 月 1 日至 6 月 12 日休婚假。

本體 Body

❶ I would greatly appreciate it if you could send your signed contract before I go on maternity leave.

如果您可以在我休產假之前將您簽好的合約寄回來，我將不勝感激。

❷ Please let me know whether you would like to schedule our meeting before or after I go on annual leave.

請告訴我您希望將我們的會面安排在我休年假前，或是我休年假後。

❸ If you need assistance during my absence, please call 1234-5678 for Molly Yang, who will cover my position while I am on leave.

若您在我休假期間需要協助，請撥打 1234-5678 找楊茉莉，她將在我休假期間代理我的職務。

❹ Please kindly be advised that my colleague, David Turner, will assume my responsibilities in my absence.

我的同事大衛拓納會在我休假期間代理我的職務。

❺ Our Deputy Director will act for me before I return to work.

我們的副主任將會在我在我回到工作崗位之前代理我的工作。

❻ My colleague, Jennifer Rover, will be in charge of your project during my absence.

在我休假期間，我的同事珍妮佛羅福將會負責您的案子。

結尾 Closing

❶ If you need urgent assistance from me, please call my cellphone 0912345678.

如果您有要事需要我協助，請撥打我手機 0912345678。

❷ **I can still be reached by my cellphone during my leave.**
我休假期間，您仍可以打手機與我聯絡。

❸ **I will get in touch with you as soon as I am back to work.**
我一回到工作崗位就會立刻與你聯絡。

❹ **Thank you for your continued support.**
感謝您一直以來的支持。

 **不可不知的實用E-mail字彙**

- **annual/yearly leave** 年假
- **maternity leave** 產假
- **personal leave** 事假
- **parental leave** 育嬰假
- **bereavement leave** 喪假
- **marriage leave** 婚假
- **act for sb.** 代理某人
- **cover sb.'s position** 代理某人職務
- **assume sb.'s responsibility** 代理某人職責

04 人事異動通知

A Staff Change Notice Letter

寫作要點 Key Points:

Step 1: 告知公司內部人事異動。

Step 2: 介紹接任的工作人員，並提供聯絡資訊。

Step 3: 懇請對方繼續合作關係，並感謝對方。

 實際E-mail範例

| 寫信 ▼ | 刪除 | 回覆 ▼ | 寄件者： Ginny |

Dear Sir/Madam,

Introduction

We are writing this letter to notify you that our sales representative Mr. David Guo is transferring to our New York Branch next month.

Body

Please kindly be informed that Ms. Linda Wang will succeed to this position and provide her best service to you.

Closing

Please feel free to contact us if you have any questions regarding this personnel change.

We thank you for your continued support and cooperation.

Best regards,
Signature of Sender
Sender's Name Printed

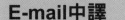

```
┌─────────────────────────────────────────────────────────────┐
│ E-mail中譯   寫信│▼   刪除   回覆│▼   寄件者：Ginny       │
├─────────────────────────────────────────────────────────────┤
```

您好：

開頭

我們特地寫這封信通知您，我們的業務代表郭大衛先生下個月即將轉調至我們的紐約分公司。

本體

在此通知您，王琳達小姐將會接任此一職務，並為您提供她最好的服務。

結尾

若您對此人事異動有任何問題，請儘管與我們聯絡。

感謝您一直以來的支持與合作。

敬祝安康，
簽名檔／署名

 各段落超實用句型

說明：畫底線部分的單字可按照個人情況自行替換

開頭 Introduction

❶ This letter serves to inform you that I am transferring to our <u>branch office</u> <u>next month</u>.

這封信是要通知您，我即將在<u>下個月</u>調到<u>分公司</u>。

❷ I am writing this letter, to inform you of my transfer.

我是寫信來通知您關於我調職的事。

❸ I, with this letter, would like to inform you that I will resign from my position, starting from <u>March 1ˢᵗ</u>.

我想以此信通知您，我將於 <u>3 月 1 日</u>離職。

❹ It is with regret that we have to announce that our <u>Director</u> <u>Mr. Brendon Smith</u> is retiring from his position <u>next month</u>.

很遺憾在此宣布，我們的<u>主任布蘭登史密斯先生</u>即將在<u>下個月</u>從他的職務退休。

❺ The purpose of this letter is to announce my resignation.

這封信的目的是要宣布我離職的消息。

本體 Body

❶ I am pleased to announce that my colleague, <u>Ms. Helen Wood</u>, will replace me as the <u>Chief Editor</u> after I retire.

我很高興在此宣布，我的同事<u>海倫伍德女士</u>，將會在我退休之後接替我擔任<u>總編輯</u>一職。

❷ Please kindly note that <u>Mr. John Ellen</u> will succeed <u>Ms. Jennifer Tien</u> to her position as our <u>General Manager</u>.

在此通知您，<u>約翰艾倫先生</u>將會接替<u>田珍妮小姐</u>，擔任本公司<u>總經理</u>一職。

❸ I am pleased to introduce <u>Mr. Joseph Liang</u>, who will be my replacement after my resignation.

跟高興向您介紹<u>梁約瑟先生</u>，他將在我離職後接替我的工作。

❹ Our new colleague, <u>Mr. Robin Tong</u>, is replacing me as <u>Director of Marketing Department</u> as of <u>October 1st</u>.

我們的新同仁<u>童羅斌先生</u>，將自<u>10月1日</u>起接替我的職務，擔任<u>行銷部主任</u>一職。

❺ I am sure you will be pleased to have <u>Mr. Peter Song</u>, my successor, to continue providing our best service to you.

我的接任者——<u>宋彼得先生</u>將繼續為您提供我們最好的服務，相信您會對於這個消息感到很高興。

❻ For future correspondences, please contact <u>Mr. Gary Lee</u>, who will take over my duties after my retirement.

未來的聯絡事宜請找<u>李蓋瑞先生</u>。我退休後，我的職務將由他負責。

結尾 Closing

❶ It's been a pleasure working with you.

和您合作很高興。

❷ I've enjoyed working with you.

很高興與您合作。

❸ I appreciate your support over the past <u>two years</u>.

感謝您過去<u>兩年</u>來的支持。

❹ I would like to take this opportunity to express my gratitude for your kind support and assistance during the past years.

我想利用這個機會，感謝您過去幾年來的支持與幫助。

❺ Please extend your support to my successor.

懇請您繼續支持我的接任者並保持合作。

❻ Working with you in the past <u>four years</u> has been extremely pleasant for me, and I wish you good luck in all your future businesses.

過去<u>四年</u>與您的合作非常愉快，祝福您未來的事業一切順利。

 # 不可不知的實用E-mail字彙

- **successor** 接任者
- **success to sb.'s position** 接任某人職務
- **take over sb.'s responsibility** 接任某人職責
- **announce sb.'s resignation** 宣布某人離職
- **announce sb.'s retirement** 宣布某人退休
- **inform of sb.'s transfer** 通知某人調職

地址、電話變更通知

A Change of Address/ Number Notice Letter

寫作要點 Key Points:

Step 1: 告知對方公司地址／電話有異動

Step 2: 請對方更改公司聯絡資訊。

Step 3: 感謝對方長期的支持，並表示期望未來能繼續提供服務。

實際E-mail範例

| 寫信┃▼ | 刪除 | 回覆┃▼ | 寄件者： | Ginny |

Dear Sir/Madam,

Introduction

This letter serves to notify you that we are moving our office to a new place on March 11th, 2017.

Body

Please change the information in your records and send future communications to the new address. Thank you very much.

Our current address:

Manor Hall, Lower Clifton Hill, Bristol, BS8 1BU

Our new address:

45 Queens Rd., Bristol, BS8 1RE

Once you update the records, please let us know by sending us a confirmation. Both our office number and email address remain unchanged.

Closing

We appreciate your continued support and look forward to providing you the best service in the future.

....

Best regards,
Signature of Sender
Sender's Name Printed

E-mail中譯　寫信▼　刪除　回覆▼　寄件者：Ginny

您好：

....

開頭

此信旨在通知您，本公司將在 2017 年 3 月 11 日遷至新址。

本體

煩請更改貴公司的通訊錄資訊，日後即以新址通信往來。非常感謝。

本公司舊址：

Manor Hall, Lower Clifton Hill, Bristol, BS8 1BU

本公司新址：

45 Queens Rd., Bristol, BS8 1RE

一旦您更新資訊，煩請來函予以確認。
本公司聯絡電話及電子郵件地址均維持不變。

結尾

感謝您繼續支持，並期望能在未來持續提供您最佳的服務。

....

敬祝安康，
簽名檔／署名

各段落超實用句型

說明：畫底線部分的單字可按照個人情況自行替換

開頭 Introduction

❶ This letter is to inform you of the relocation of our head office on December 1st, 2017.

謹以此郵件通知您，本公司總部將於 2017 年 12 月 1 日遷移至新址。

❷ We are happy to inform you that our office will be relocated on the eighth of this month.

很高興通知您，本公司將於本月 8 日遷移至新址。

❸ It is a great pleasure to announce the relocation of our Manchester Branch on December 1st, 2017.

很高興在此宣佈本公司曼徹斯特分部將於 2017 年 12 月 1 日遷移至新址。

❹ I am writing this letter to inform you that our office number has been changed to 2213-0000.

我寫這封信是要通知您，我們辦公室的電話已經改成 2213-0000。

❺ This letter serves to inform you that my extension number will be changed to 8372 after the relocation of my office.

這封信是要通知您，在辦公室搬遷後，我的分機號碼將會改成 8372。

本體 Body

❶ Please update your records with our new address accordingly.

請將我們變更的新址更新至您的通訊錄。

❷ To offer better service to our growing customers, we will be moving our business location.

為了提供更優質的服務予日益增加的顧客群，本公司將遷移營業位址。

❸ Our last day of operations at the current location is March 31st, 2017.

2017 年 3 月 31 日將會是在目前地址營業的最後一日。

❹ We will be ready to accept shipments from you at our new location on April 1st, 2017.

2017 年 4 曰 1 日起新據點將可開始收發貨。

❺ **The following are our new office phone number and address:**
以下是我們新辦公室的電話及地址：

❻ **Both our office phone number and fax number remain the same.**
我們的辦公室電話及傳真號碼均維持不變。

結尾 Closing

❶ **We look forward to continuing to do business with you.**
我們期待能繼續與您合作。

❷ **We will strive for our best performance and efficiency at the new office.**
我們將在新的辦公地點力求最佳工作表現及效率。

❸ **We will continue providing our services to you at our new office.**
我們將在新辦公室繼續為您提供服務。

❹ **Thank you for your continued support and cooperation.**
感謝您持續支持與合作。

 # 不可不知的實用E-mail字彙

- **update your records** 更新您的通訊錄
- **office relocation** 辦公室遷址
- **office number** 辦公室電話
- **office address** 辦公室地址
- **current location** 現址
- **office fax number** 辦公室傳真電話
- **extension number** 分機號碼
- **remain the same** 維持不變

A Meeting Minutes Letter

寫作要點 Key Points:

Step 1: 感謝大家與會。
Step 2: 提供會議紀錄，請大家閱覽。
Step 3: 歡迎大家提出問題，並期待下次開會。

實際E-mail範例

寫信 ▼	刪除	回覆 ▼	寄件者： Ginny

Dear Sir/Madam,

Introduction

Thank you all for another great meeting today. I am proud of all the accomplishments we have made in the meeting and would like to invite you to review the meeting minutes.

Body

The minutes of the meeting have been transcribed and proofread and are now attached for your review.

Please do not hesitate to contact me if you would like to discuss any topics contained in the minutes.

Closing

Look forward to our next meeting in July.

Best regards,
Signature of Sender
Sender's Name Printed

E-mail中譯

寫信 ▼　刪除　回覆 ▼　寄件者：Ginny

您好：

開頭

感謝所有人在今天早上又開了一次美好的會議。我對於我們在會議中所討論完成的事項引以為傲，並希望能邀請您詳閱這次的會議紀錄。

本體

這份會議紀錄已經被謄寫並校對過，而且現在以附件方式這封信裡供您詳閱。

如果您想討論紀錄中任何議題，請直接與我聯絡。

結尾

期待我們 7 月份的會議。

敬祝安康，
簽名檔／署名

各段落超實用句型

說明：畫底線部分的單字可按照個人情況自行替換

開頭 Introduction

❶ It was great meeting you all at the monthly Department Meeting held last Friday, August 12ᵗʰ.

很高興在上週五，8 月 12 日舉行的部門月會中見到各位。

❷ Glad we got to meet at the Department Meeting held yesterday morning.

很高興我們在昨天上午舉行的部門會議中見面。

❸ I am excited about the progress we have made in the meeting.

我對於我們在會議中的進展感到非常興奮。

❹ Please take a few minutes to review the meeting minutes to see what we have accomplished.

請花幾分鐘的時間閱讀一下會議紀錄，看看我們完成了哪些事項。

⑤ I would like to thank you all for another great meeting <u>this morning</u>.
我想感謝大家，<u>今天早上</u>又開了一次很棒的會議。

⑥ I am glad to see that we have had such a lucrative meeting <u>this Monday</u>.
很開心看到我們<u>本週一</u>的會議收穫非常多。

本體 Body

❶ Attached are the detailed minutes of our meeting held <u>this morning</u>.
附件為<u>今天上午</u>舉行的會議詳細紀錄。

❷ Please find attached the full transcript of the meeting minutes.
完整的會議紀錄請見附件。

❸ The meeting minutes have been compiled, transcribed, and proofread, and are now attached to this letter for your review.
會議紀錄已經編整、紀錄並校對過，並且隨信附件以供您閱覽。

❹ Please feel free to contact me if you notice any discrepancies between the minutes and the actual meeting content.
若您發現紀錄與實際會議內容有任何出入，請儘管與我聯絡。

❺ Please contact me if you would like to discuss any topic contained in the minutes.
若您想討論任何紀錄中所包含的議題，請與我聯絡。

❻ In the <u>monthly Department meeting</u> of <u>July</u>, we discussed all the upcoming projects for the next <u>six months</u>.
我們在 <u>7 月份</u>的<u>部門月會</u>中，討論了未來<u>六個月</u>即將執行的專案。

❼ Please note that these minutes are the official minutes of the meeting.
請注意，這份紀錄是正式的會議紀錄。

❽ The official minutes of the <u>committee meeting</u> have been posted on <u>the bulletin board beside the elevator</u> for your viewing pleasure.
<u>委員會會議</u>的正式紀錄都已經貼在<u>電梯旁的公佈欄</u>上供您觀看了。

結尾 Closing

❶ **I look forward to seeing you again at the next meeting.**
期待能在下次的會議中再見到各位。

❷ **Look forward to our next monthly meeting.**
期待下一次月會。

❸ **The next meeting will be held in the main conference room at 1:30 pm on November 13th, 2017.**
下次會議將會在 2017 年 11 月 13 日下午 1 點 30 分，於主會議室舉行。

❹ **I look forward to brainstorming with you all in next Monday's meeting.**
我很期待下週一的會議上也能與大家一起腦力激盪。

 # 不可不知的實用E-mail字彙

- **meeting minutes** 會議紀錄
- **meeting transcript** 會議紀錄
- **official minutes of the meeting** 正式會議紀錄
- **draft of meeting minutes** 會議紀錄草稿
- **meeting minute taker** 會議紀錄者
- **meeting attendee** 出席會議者
- **moderator** 會議主席

Part8

進度催促篇

01 催促出貨

A Shipment Request Letter

寫作要點 Key Points:

Step 1: 說明來信的目的與某訂單有關。
Step 2: 提醒對方應交貨日期,並表示對方已延遲出貨。
Step 3: 催促對方立刻出貨,並告知對方對應延遲出貨的處理方式。

實際E-mail範例

寫信 | ▼　　刪除　　回覆 | ▼　　　寄件者: Ginny

Dear Sir/Madam,

Introduction

I am writing this letter with regard to the order for 20 electric water boilers that we placed with you on April 30th, 2017.

Body

According to our agreement, our order should be shipped within five working days once our payment is settled. That is to say, we should have received our order by last Friday. However, our commodities haven't been delivered to us so far.

Closing

Should you fail to deliver our order by the end of this week, we will have no option but to exercise our right to terminate our agreement with you and seek compensation from your company.

Best regards,
Signature of Sender
Sender's Name Printed

E-mail中譯　寫信┃▼　刪除　回覆┃▼　寄件者：Ginny

您好：

> **開頭**
>
> 我寫的這封信是有關於我們在 2017 年 4 月 30 日向您訂購的 20 台電動熱水瓶的訂單。
>
> **本體**
>
> 根據我們的合作協議，我們的訂單應該在帳款支付之後五個工作天內出貨。也就是說，我們在上週五就應該要收到我們訂購的物品。然而，我們訂購的商品一直到現在都還沒有送達。
>
> **結尾**
>
> 如果您未能在本週結束前幫我們出貨，我們將不得不行使我們的權利，終止與您的合作協議，並向貴公司要求賠償。

敬祝安康，
簽名檔／署名

各段落超實用句型

說明：畫底線部分的單字可按照個人情況自行替換

開頭 Introduction

❶ Our order is supposed to be delivered to us <u>two weeks</u> ago.
我們訂購的商品在<u>兩週</u>前就應該寄來給我們了。

❷ I am writing this letter to inform you of the delay of delivery of <u>500 umbrellas</u> that we ordered from you.
我寫這封信來通知您，我們向貴公司訂購的 <u>500 支雨傘</u>交貨延遲了。

❸ We regret to point out that we have not yet received our order, which should be shipped to us <u>seven days</u> ago.
我們必須很遺憾地告訴您，我們還沒有收到<u>七天</u>前就應該送來的訂單貨品。

❹ You accepted our order that we placed on September 15th and claimed to ship our order to us by September 22nd.

您受理了我們 9 月 15 日下的訂單，並聲稱會在 9 月 22 日前出貨。

❺ I am writing this letter to let you know that our order has not yet reached us.

寫這封信是要讓您知道，我們訂購的商品到現在還沒送來。

本體 Body

❶ It should have arrived two days ago.

兩天前就該到了。

❷ According to our contract, our order should be shipped within two weeks after the order was accepted.

根據我們的合約，我們的訂單應該要在受理訂單後兩週內出貨。

❸ It has been three weeks since the payment of my order was settled in full, however, I still haven't received my commodit my ies.

我的訂單帳款三週前就已全額結清，然而我卻還沒有收到商品。

❹ You agreed to have our order shipped to us by September 30th, but we have not received it up to this date.

您同意在 9 月 30 日前將我們的訂單出貨，但我們至今仍未收到物品。

❺ I was surprised to find that my order hasn't arrived yet, even though you agreed to ship it a month ago.

我很驚訝地發現，雖然您同意一個月前就要送出我的訂貨，我卻現在都還沒收到。

結尾 Closing

❶ Thank you for your prompt attention to this matter.

感謝您立即處理這個問題。

❷ We must ask you to have our order shipped to us as soon as possible.

我們必須要求貴公司立即幫我們的訂單出貨。

❸ If you cannot proceed with the shipment immediately, we would have to cancel our order.

若您無法立刻出貨，我們將必須取消訂單。

❹ The delay of the delivery has been causing considerable inconvenience to our company.

延遲出貨為我們公司帶來相當大的困擾。

❺ Please understand that this delay has caused us great loss in business.

請您明白這次延遲出貨已經造成我們業務上極大損失。

❻ We would terminate our contract with you if this happens again.

若這種事情再次發生，我們將會終止與貴公司的合約。

 不可不知的實用E-mail字彙

- **delay of delivery** 延遲出貨
- **loss in business** 業務上的損失
- **terminate the contract** 終止合約
- **cancel the order** 取消訂單
- **seek compensation** 尋求賠償

A Return-of-Original-Document Request Letter

寫作要點 Key Points:

Step 1: 說明來信目的乃有關出借之資料或樣品。
Step 2: 提醒對方返還物品，並提出具體返還時間。
Step 3: 請對方立即回應此問題。

 實際E-mail範例

| 寫信 ▼ | 刪除 | 回覆 ▼ | 寄件者： | Ginny |

Dear Sir/Madam,

Introduction

I am writing this letter with regard to the project planning files that you borrowed to make copies of this past Monday.

Body

We are going to need the files to discuss our project plans further in detail in our weekly department meeting, scheduled to be held this Friday morning. Therefore I would like to request you to kindly return the files to our department as soon as possible.

Closing

I understand you may need some more time to make the copies you need, but please have them returned to us by this Thursday at the latest.

I appreciate your understanding and cooperation.

Best regards,
Signature of Sender
Sender's Name Printed

E-mail中譯　寫信 ▼ ｜刪除 ｜回覆 ▼ ｜寄件者：Ginny

您好：

開頭

我是為了您在星期一借去影印的專案企劃檔案而寫這封信的。

本體

我們將會需要這些檔案，以便於本週五上午舉行的部門週會中進一步詳細討論這些專案計劃。因此我想請您盡快將檔案歸還我們部門。

結尾

我知道您可能需要多一些時間來影印您需要的東西，但請務必最晚在這星期四之前還給我們。

感謝您的諒解與配合。

敬祝安康，
簽名檔／署名

各段落超實用句型

説明：畫底線部分的單字可按照個人情況自行替換

開頭 Introduction

❶ With reference to the <u>sample products</u> you borrowed from our department <u>last week</u>, ...

有關您<u>上週</u>向本部門借的<u>產品樣本</u>，……

❷ With regard to the <u>project report documents</u> you borrowed <u>yesterday</u>, ...

有關您<u>昨天</u>借閱的<u>專案報告文件</u>，……

❸ I am writing this letter regarding the <u>graphic design portfolio</u> that you borrowed from us <u>two months</u> ago.

我是為您<u>兩個月</u>前向我們借閱的<u>平面設計圖集</u>而寫這封信。

❹ **Thanks for accepting our invitation to try out our new product.**
感謝您接受本公司邀請試用我們的新產品。

❺ **We hope you have tried out our products and find them satisfactory.**
我們希望您已經試用過我們的產品，並對它們很滿意。

本體 Body

❶ **Please return the sample products at your earliest convenience.**
麻煩您盡快在方便的時間返還產品樣本。

❷ **Please let us know when the best time for us to retrieve our sample products from you is.**
請讓我們知道何時可以向您取回試用產品。

❸ **We would like to request you to send our sample products back to our office.**
我們想要請您將產品樣本送回我們公司。

❹ **We would greatly appreciate it if you could return our graphic design portfolio as soon as possible.**
如您能盡快返還我們的平面設計圖集，我們將十分感謝。

❺ **I would like to ask you to return the personnel files you borrowed yesterday.**
我想請您返還昨天借的人事檔案。

❻ **I would appreciate it if you could return it by the end of this week.**
如果您能在本週結束前返還，我將不勝感激。

❼ **I understand that you may need these files, but we cannot do without them in next week's Monthly Meeting.**
我瞭解您也許需要這些資料，但我們下週的月會必須要用到它們。

結尾 Closing

❶ **Please let me know when you could return the documents at the earliest.**
請讓我知道您最早何時可以返還這些文件。

❷ **Please keep me informed if you need more time to inspect those files.**

若您需要更多時間檢閱那些資料，請通知我。

❸ **We look forward to your positive reply.**

期待您的正面回應。

❹ **Thank you in advance.**

在此先向您道謝。

❺ **Please return <u>the documents</u> to my desk <u>tomorrow</u> at the latest.**

請最晚在<u>明天</u>前將<u>文件</u>歸還到我的桌上。

不可不知的實用E-mail字彙

- **borrow sth. for inspection** 借某物去看
- **borrow sth. to make copies** 借某物去影印
- **try out the samples** 試用樣品
- **give back** 歸還
- **return sth. in person** 親自歸還
- **retrieve the samples** 取回樣品

03 催促提供發票

An Receipt Request Letter

寫作要點 Key Points:

Step 1: 告知對方未提供發票。
Step 2: 催促對方提供發票。
Step 3: 請求對方立即回應此問題。

 實際E-mail範例

寫信 ▼   刪除   回覆 ▼   寄件者： Ginny

Dear Sir/Madam,

Introduction

Thank you for providing your painting and decorating services for our office building. I am pleased to inform you that we have settled our payment for your services. However, we must remind you that we haven't received the receipt from you.

Body

Please settle the receipt for our payment at your earliest convenience because we will need the receipt to request for reimbursement.

Closing

Thank you very much for your prompt attention to this matter.

Look forward to hearing from you soon.

Best regards,
Signature of Sender
Sender's Name Printed

E-mail中譯　　寫信│▼　　刪除　　回覆│▼　　寄件者：Ginny

您好：

開頭

感謝您為本公司辦公大樓所提供的粉刷及裝修工程。我很高興通知您，我們已經支付了貴公司的工程款。然而，我們必須提醒您，我們還沒有收到貴公司的發票。

本體

麻煩您盡快幫我們的帳款開立發票，因為我們將需要發票來請款報帳。

結尾

期待得到您的回覆。感謝您對此事的立即關注。

希望能很快得到您的回覆。

敬祝安康，
簽名檔／署名

各段落超實用句型

說明：畫底線部分的單字可按照個人情況自行替換

開頭 Introduction

❶ **Our order has arrived this morning. Thank you very much.**

我們訂購的物品今天上午已經送達了。非常感謝您。

❷ **This letter is to inform you that we have settled our payment for your window cleaning service.**

謹以此信通知您，我們已經支付您清洗窗戶的服務費用。

❸ **I am pleased to notify you that the payment of our order for 100 thermometers has been settled.**

我們很高興通知您，我們訂購 100 個熱水瓶的費用已支付完成。

❹ We have received our order, but we found that the receipt didn't come with our shipment.

我們已經收到訂購的產品，但是發現發票沒有跟著寄過來。

本體 Body ....

❶ I am writing to remind you to issue a receipt for our purchase.

我寫信來提醒您，為我們購買的產品開立發票。

❷ The purpose of this letter is to request the receipt for our order.

這封信的目的是要索取我們訂單的發票。

❸ We must inform you that we haven't got the receipt from you.

我們必須通知您，我們尚未收到貴公司開立的發票。

❹ We are still waiting for the receipt for our payment.

我們還在等我們帳款的發票。

❺ You neglected to settle the receipt for our order purchased from you.

您忘了為我們向您購買的產品開立發票。

❻ I'd like to remind you that I haven't received the receipt for the purchase I made with you last month yet.

我想提醒您，我到現在還沒收到上個月與您購物的收據。

結尾 Closing ....

❶ Please issue a receipt for our payment at your earliest convenience.

請您盡快為我們的帳款開具發票。

❷ We need you to issue a receipt for us without delay.

我們需要您立刻開發票給我們。

❸ We must request you to make out a receipt for us A.S.A.P.

我們必須請你們立刻開具發票給我們。

❹ We would greatly appreciate it if you could settle the receipt for our payment immediately.

若您能立刻為我們的帳款開立發票，我們將不勝感激。

❺ **Please issue a receipt for our payment and have it sent to us no later than June 30ᵗʰ.**

請為我們的帳款開立發票,並在 6 月 30 日之前交寄給我們。

❻ **We need the receipt by the end of this month for tax purposes.**

我們需要在本月底之前拿到發票,以便報稅使用。

❼ **I am in urgent need of the receipt and would really appreciate it if you could send it to me without delay.**

我非常需要這張收據,若您能將它立即寄送給我,我將感激不盡。

 不可不知的實用E-mail字彙

- **request a receipt** 索取發票
- **settle a receipt** 開立發票
- **issue a receipt** 開立發票
- **make out a receipt** 開立發票
- **put off writing a receipt** 拖延開立發票
- **issue receipt lottery** 開立統一發票
- **request for reimbursement** 報帳請款

A Late Payment Request Letter

 實際E-mail範例

| 寫信 ▼ | 刪除 | 回覆 ▼ | 寄件者： Ginny |

Dear Sir/Madam,

Introduction

We are writing this letter regarding your order No. 12345 for 300 lightweight jackets. Your order was shipped to you two weeks ago and it has come to our attention that your payment is overdue.

Body

We would like to request you to make the payment in full within the next three business days.

Closing

Should we not receive your payment by the expected date, we will be left with no alternative but to take legal actions.

Thank you for your immediate attention to this matter.

Best regards,
Signature of Sender
Sender's Name Printed

E-mail中譯　　[寫信|▼]　[刪除]　[回覆|▼]　寄件者：[Ginny]

您好：

開頭

這封信是有關您訂購 300 件輕量級外套，編號 12345 的訂單。您所訂購的商品已經在兩週前寄送給您了，但是我們注意到您的帳款逾期未繳。

本體

我們想請您在未來三個營業日之內全額結清帳款。

結尾

假如我們沒有在預定日期前收到您的帳款，我們將不得不採取法律行動。

感謝您對此事的立即關注。

敬祝安康，
簽名檔／署名

 ## 各段落超實用句型

說明：畫底線部分的單字可按照個人情況自行替換

開頭 Introduction

❶ **It has come to our attention that your payment is overdue.**
我們注意到您的帳款尚未支付。

❷ **I am writing to inform you of your late payment.**
我是寫信來通知您帳款延遲了。

❸ **We haven't received your payment, which was due <u>two weeks</u> ago.**
我們還沒有收到您<u>兩週</u>之前就應繳付的帳款。

❹ **We would like to remind you that your payment is due on <u>June 30th, 2017</u>.**
我們想提醒您，您的帳款在 <u>2017 年 6 月 30 日</u>就應該支付了。

❺ We must remind you that your payment is already due.

我們必須提醒您，您的帳款已經到期了。

❻ You are expected to settle the payment within <u>seven days</u> of receipt of your order.

您應該要在收貨<u>七天</u>內支付帳款。

❼ This message is to remind you of the outstanding payment for your order #12411.

這則訊息是要提醒您，您的訂單 <u>12411</u> 號目前還有未付款項。

本體 Body

❶ We would like to request you to complete the payment within the next <u>five business days</u>.

我們希望您能在未來<u>五個工作天</u>之內結清帳款。

❷ We must ask you to make the payment in full settlement within <u>two days</u>.

我們必須請您在<u>兩天</u>內繳付全額帳款。

❸ Attached is the copy of your receipt that is due on receipt.

附件是您收到就必須支付的收據副本。

❹ Please forward us the payment in full at your earliest convenience.

請盡快將帳款全額轉給我們。

❺ We would be highly grateful to receive your remittance as soon as possible.

若我們能盡快收到您的匯款，將不勝感激。

❻ We expect to receive your payment no later than <u>May 31st</u>.

我們希望能在 <u>5 月 31 日</u>之前收到您的貨款。

❼ The payment must be complete within <u>the next week</u>.

<u>下週</u>內必須要完成付款。

結尾 Closing

❶ Please give your prompt attention but to this matter.

請您立刻關注這件事。

❷ We would have no alternative but to take legal actions if we do not receive payment by the expected date.

如果我們未能在預定日期之前收到帳款，將不得不採取法律行動。

❸ Please note that you will be charged a <u>5%</u> interest charge for any outstanding balance greater than <u>30 days</u>.

請注意，若是超過<u>三十天</u>有任何未支付款項，您將必須按月繳納<u>百分之五</u>的利息。

❹ If we continue to not hear from you, we will have no choice but to take legal actions.

若您依舊不回覆我們，我們只好尋求法律途徑了。

 不可不知的實用E-mail字彙

• **late payment** 延遲付款
• **overdue payment** 過期帳款
• **complete the payment** 結清帳款
• **make the payment in full settlement** 全額繳付帳款
• **charge interest charge** 收取利息費用

A Signed-Contract Request Letter

實際E-mail範例

寫信 ▼	刪除	回覆 ▼		寄件者： Ginny

Dear Sir/Madam,

Introduction

In our meeting with you last Friday, we agreed with all the terms and conditions, and we have modified the terms regarding delivery date in the contract as you requested.

Body

Attached is the modified contract, please review the terms and conditions again and send us the signed contract no later than this Friday so that we can put your order into production right away.

Closing

Please allow me to remind you that any delay of the contract will impede our production procedure.

Thank you very much for your prompt attention to this matter.

Look forward to hearing from you soon.

Best regards,
Signature of Sender
Sender's Name Printed

E-mail中譯

寫信 ▼　刪除　回覆 ▼　寄件者：Ginny

您好：

開頭

在我們上週五與您的會議中，我們已經在所有條約與條款方面取得共識了。

而且我們已經如您所要求，修改了有關交貨日期的合約內容。

本體

附件是已經修改好的合約。請您再次閱讀合約內容與條件，並將簽好的合約於本週五前寄給我們，好讓我們能立刻將您的訂單交付生產。

結尾

請容我提醒您，合約的延誤會阻礙我們的生產進度。

感謝您對此事的立即關注。

期待能很快得到您的回覆。

敬祝安康，
簽名檔／署名

各段落超實用句型

說明：畫底線部分的單字可按照個人情況自行替換

開頭 Introduction

❶ **We herewith attach the modified contract.**

我們隨此信附上修改過的合約。

❷ **Both parties have agreed with all the terms and conditions in the negotiation meeting.**

在協商會議中，雙方已經同意所有條件與條款。

❸ **We have modified the terms regarding** <u>delivery date</u> **as you requested.**

我們已經照您所要求，修改了有關交貨日期的合約內容。

❹ We are glad that we have finally come to an agreement on all terms and conditions in our negotiation meeting.

很高興在協商會議中，我們終於對所有條件與條款取得了共識。

❺ I hope you have reviewed the modified contract and found all terms and conditions agreeable.

我希望您們已經看過修改後的合約，並對所有條件與條款都無異議。

本體 Body

❶ Please make sure you send the signed contract to us no later than May 1st.

請務必在 5 月 1 日之前將簽好的合約寄給我們。

❷ We need to receive your signed contract to put your order into production.

我們需要收到您簽好的合約，才能將您的訂單交付生產。

❸ We must remind you that any delay of the contract will impede our production procedure.

我們必須提醒您，合約有任何延誤都會阻礙生產進度。

❹ We are waiting for the signed copies of the agreement to commence construction.

我們在等您簽好的合約副本，以便開始施工。

❺ Please allow me to remind you that if you don't have the contract signed within this week, the production of your order will be delayed.

請容許我提醒您，如果您未能在本週內簽約，將會延誤您的訂單生產。

❻ Please understand that our services will not begin until we have received your signed contract.

請瞭解，若沒有收到您簽好的合約，我們就無法開始提供服務。

結尾 Closing

❶ We expect to hear from you no later than this Friday.

我們期望能在這週五之前得到您的回覆。

❷ Thank you for your immediate attention to this matter.

感謝您對此事的立即關注。

❸ **Please contact me immediately if you have any questions regarding the terms of the contract.**

如果您對條約內容有任何問題，請立刻與我聯絡。

❹ **Should you have any other objection to the terms, please contact us at once.**

假如您對條約內容有任何異議，請立刻與我們聯絡。

❺ **Thank you again for this opportunity to work with you.**

再次感謝您給我們與您合作的機會。

❻ **To avoid any further delays, please return the signed contract as soon as possible so that the project can be finished on time.**

為避免更多延遲，請盡快將簽好的合約返還給我們，這樣案子才能在時限內完成。

 不可不知的**實用E-mail字彙**

• **modified contract** 修改過的合約
• **signed contract** 簽好的合約
• **come to agreement** 取得共識
• **terms and conditions** 條件與條款
• **objection** 異議

Part 9

感謝與道歉篇

A Late Notice-of-Absence Letter

 實際 E-mail 範例

| 寫信 ▼ | 刪除 | 回覆 ▼ | 寄件者： | Ginny |

Dear Sir/Madam,

Introduction

It is with regret that I must notify you that I will not be able to attend our department meeting this afternoon.

Body

Due to sudden and unexpected circumstances, I will not be in the office for the next three days.

Closing

I sincerely apologize for the late notice.

Best regards,
Signature of Sender
Sender's Name Printed

E-mail中譯　　寫信｜▼　　刪除　　回覆｜▼　　寄件者：Ginny

您好：

....

開頭

很抱歉必須通知您，我將無法出席今天下午的部門會議。

本體

由於臨時發生了一點突發狀況，未來三天我將不會進公司。

結尾

我誠摯為發出這樣的臨時通知，向您道歉。

....

敬祝安康，
簽名檔／署名

各段落超實用句型

說明：畫底線部分的單字可按照個人情況自行替換

開頭 Introduction ....

❶ **I am writing this letter with reference to the conference scheduled 3 pm today.**

為了今天下午 3 點的會議，寫這封信給大家。

❷ **It is with great regret that I have to inform you that we will not be able to attend the meeting scheduled 10 am this morning.**

很抱歉通知各位，我們將無法出席今天上午十點的會議。

❸ **With respect to today's luncheon with our clients, ...**

有關今日與客戶的午餐會議，……

❹ **The purpose of this letter is to inform you that we cannot make it to this afternoon's meeting.**

這封信的目的是要通知各位，我們無法出席今天下午的會議。

⑤ I with regret have to inform you that I will be absent from today's meeting.

很抱歉通知您，我無法出席今天的會議。

⑥ I am very sorry to inform you on such short notice, but I will not be able to make it to the 2 pm meeting.

很抱歉這麼晚才通知您，但下午兩點的會議我無法參加。

本體 Body

❶ We are regretful to inform you of our absence at the last minute.

我們很抱歉臨時告知無法出席。

❷ In fact, I will take the next three days off because of urgent family issues.

事實上，我因為緊急的家庭因素，未來三天將會請假。

❸ I am unable to attend tomorrow's meeting because of urgent personal errands.

因為緊急個人因素，我將無法參加明天的會議。

❹ I couldn't, under the circumstances, attend Thursday's seminar.

在這種情況下，我無法參加星期四的研討會。

❺ Due to sudden and unexpected circumstances, we're afraid that we will be absent from this afternoon's weekly meeting.

由於發生了一點臨時狀況，今天下午的週會我們恐怕得缺席。

❻ An emergency has happened and I will be unable to make it to the office building in time.

發生了一點急事，我來不及回到公司大樓。

結尾 Closing

❶ I am sorry for the late notice.

我對臨時通知感到很抱歉。

❷ I apologize for all the inconveniences that my absence may have caused.

我對於缺席會議所可能造成的不便，向您道歉。

❸ **Please accept our apology for our absence.**
請接受我們對無法出席的道歉。

❹ **We sincerely apologize for giving you the notice at the last minute.**
我們對臨時通知您，感到萬分抱歉。

❺ **I feel very sorry about not being able to attend the conference.**
我對無法出席會議感到非常抱歉。

❻ **I am deeply sorry for this sudden development, and will do what I can to make up for it after I return.**
對於此突然的發展，我非常抱歉，將會在回來後盡力補償。

 # 不可不知的實用E-mail字彙

• **late notice** 臨時通知
• **at the last minute** 臨時
• **due to sudden and unexpected circumstances** 由於突發狀況
• **family issues** 家庭因素
• **urgent personal errands** 緊急的私人事務
• **absence from the meeting** 會議缺席

02 延遲出貨道歉

A Late Delivery Apology Letter

 實際E-mail範例

寫信 ▼	刪除	回覆 ▼	寄件者: Ginny

Dear Sir/Madam,

Introduction

With reference to your order for 1,000 down jackets that you placed with us on June 10th, we regret to have delayed the delivery.

Body

An unexpected problem occurred in our delivery process, but we have managed to fix it this morning and shipped your order immediately.

Your commodities should arrive at your appointed address this afternoon.

Closing

Please accept our sincere apology for the inconvenience that this delay has caused you.

We assure you that this kind of mistake will not happen again in our future cooperation.

Best regards,
Signature of Sender
Sender's Name Printed

E-mail中譯　　寫信｜▼　刪除　回覆｜▼　寄件者：Ginny

您好：

開頭

有關您在 6 月 10 日跟我們下的 1000 件羽絨外套訂單，我們很抱歉延遲了交貨時間。

本體

我們的出貨流程突然發生了一點問題，但是我們已經在今天上午設法解決，並立刻為您出貨。您的商品應該在今天下午就會送達您指定的地址。

結尾

對於這次延遲交貨為您帶來的不便，請接受我們的誠摯歉意。

我們向您保證，在我們未來的合作中將不會再發生相同的問題。

敬祝安康，
簽名檔／署名

各段落超實用句型

說明：畫底線部分的單字可按照個人情況自行替換

開頭 Introduction

❶ We are regretful for delaying your order.

我們對延誤您的訂單感到很抱歉。

❷ With this letter, we would like to express our regret for the late delivery of your order.

我們希望藉由此信向您表達延遲出貨的歉意。

❸ We regret to inform you that we delayed your delivery.

很抱歉通知您，我們延遲了交貨時間。

❹ This letter is to apologize for delaying your shipment.

這封信目的是為延遲出貨而向您致歉。

❺ On behalf of my company, I must apologize for the delay in delivering your order.

我謹代表本公司，為延遲寄送您的訂單商品致歉。

本體 Body ....

❶ Our supplier was unable to provide us sufficient <u>computer parts</u> that we need.

我們的供應商無法提供我們足夠的<u>電腦零件</u>。

❷ An unexpected problem occurred in our <u>delivery</u> process, but we have managed to fix it <u>this morning</u> and shipped your order immediately.

我們的<u>出貨</u>流程突然發生了一點問題，但我們已經在<u>今天上午</u>設法解決，並立刻為您出貨了。

❸ <u>Typhoon Maggie</u> has seriously disrupted our <u>production</u> schedule.

梅姬颱風嚴重地打亂了我們的<u>生產</u>進度。

❹ The unusually large number of orders during <u>the Chinese New Year Holidays</u> disordered our <u>delivery</u> schedule.

<u>農曆春節假期</u>的龐大訂單，打亂了我們的<u>配送</u>時間表。

❺ We will ship your order by fast courier immediately. You should be able to receive your commodities within the next <u>two days</u>.

我們將立刻以快遞出貨給您。您應該在<u>兩天</u>內就可以收到商品了。

❻ We have adopted a new delivery system to improve our delivery process. Therefore this kind of delay shall never happen again.

我們已經採用新的出貨系統來改善我們的出貨流程。因此這樣的延遲出貨狀況未來將不會再發生。

❼ <u>A company emergency</u> has prevented us from completing the order as scheduled.

由於<u>公司發生緊急事件</u>，我們無法如期完成訂單。

結尾 Closing ....

❶ We apologize for all the trouble that the late delivery has caused.

我們為延遲交貨所造成的一切困擾，至上歉意。

❷ **Please accept our sincere apology for the inconvenience that this delay has caused you.**
請接受我們為這次延遲交貨所為您造成的不便，致上十二萬分的歉意。

❸ **We appreciate your patience and your kind understanding.**
我們感謝您的耐心及善意體諒。

❹ **We assure you that this kind of mistake will not happen again in our future cooperation.**
我們向您保證，在我們未來的合作中類似的問題將不會再發生。

❺ **We are deeply regretful to have caused you such an inconvenience, and want to assure you that something like this will never happen again.**
造成您的不便，我們非常遺憾，希望您能明白這樣的事絕對不會再發生了。

 # 不可不知的實用E-mail字彙

- **late delivery** 延遲出貨
- **delay the delivery of order** 延遲訂單出貨
- **unexpected problem** 突發問題
- **delivery process** 出貨流程
- **delivery system** 出貨系統
- **delivery schedule** 出貨時間表，配送時間表

03 貨物瑕疵、毀損道歉

An Apology Letter for Defective/Damaged goods

寫作要點 Key Points:

Step 1: 對商品瑕疵或損毀道歉。
Step 2: 提供具體解決方法。
Step 3: 表示希望未來仍有合作機會。

實際E-mail範例

寫信 ▼　　刪除　　回覆 ▼　　　寄件者： Ginny

Dear Sir/Madam,

Introduction

Regarding your order for clothes and accessories placed with us on August 3rd, we are sorry to learn that your commodities arrived defective.

We should have double-checked the items before we shipped them to you. We hereby express our sincere apology.

Body

Due to the error on our part, we will take the responsibility of replacing your goods. Please kindly note that our courier will send you the replacements and retrieve the defective goods no later than this Friday, August 14th.

Closing

Please accept our sincere apology for this unintentional mistake.

We look forward to continue providing our best service to you in the future.

Best regards,
Signature of Sender
Sender's Name Printed

E-mail中譯　寫信▼　刪除　回覆▼　寄件者：Ginny

您好：

開頭

有關您在 8 月 3 日向本公司訂購的服飾產品，很遺憾得知您的商品寄達時是有瑕疵的。

我們應該要在寄送之前再次檢查商品。在此向您致上誠摯的歉意。

本體

由於此次是我方疏失，本公司將為您換貨以示負責。在此通知您，本公司將會在本週五，即 8 月 14 日前，將更換商品快遞寄送給您，並取回瑕疵商品。

結尾

請接受我們對這次無心之過的誠摯道歉。

我們期待未來還能繼續為您提供我們的最佳服務。

敬祝安康，
簽名檔／署名

各段落超實用句型

説明：畫底線部分的單字可按照個人情況自行替換

開頭 Introduction

❶ **We apologize for sending you a defective product.**

我們為寄給您有瑕疵的產品致歉。

❷ **We regret to learn that your order was damaged during transportation.**

我們很遺憾得知您訂購的產品在運送過程中遭到損毀。

❸ **We regret that your commodities arrived damaged.**

很遺憾您的商品送達時，是損毀的狀態。

❹ **We are sorry to know that your goods arrived with damage.**

得知您的商品送達時呈現損毀狀態，我們感到很抱歉。

❺ I, on behalf of our company, apologize for damaging your order in transit.

我謹代表本公司，為運送途中損毀您訂購的商品向您致歉。

❻ We are terribly sorry to hear that the products your received were not in their ideal condition.

很遺憾聽到您收到的產品品質並不理想。

本體 Body

❶ We will send you a replacement within the next three days.

我們會在三天內寄送新品給您。

❷ We will take the responsibility and exchange your goods.

我們會承擔責任，並為您更換貨品。

❸ Our courier will send you a replacement and retrieve the defective goods by Friday.

本公司的快遞會在週五前將更換的商品送去給您，並取回瑕疵商品。

❹ If you would like to get a full refund, we understand and respect your decision.

如果您希望全額退費，我們會理解並尊重您的決定。

❺ We will surely replace your order and have it delivered to you within two days.

我們一定會為您換貨，並會在兩天內出貨給您。

❻ This is entirely our fault. Please allow us to remedy the situation by sending you a replacement the very next morning.

這完全是我們的疏失，請允許我們明天一早寄送替代品以茲補償。

結尾 Closing

❶ We apologize for all the inconvenience.

我們為所造成的不便致上歉意。

❷ We assure you that this kind of mistake will not happen again.

我們向您保證不會再發生這類失誤。

❸ **We sincerely wish to continue providing our service to you in the future.**

我們誠摯希望將來能繼續為您服務。

❹ **Thank you for your continued patronage.**

感謝您繼續惠顧。

❺ **We do appreciate your business, and hope you will continue ordering products from us.**

我們非常感謝您的惠顧，希望您繼續向我們訂購商品。

❻ **We will be extra careful in the future to make sure that something like this will never happen again.**

我們未來將會加倍小心，確保這樣的事不再發生。

 # 不可不知的實用E-mail字彙

- **arrive with damage** 送達時已有損毀
- **in transit** 運送途中
- **during the transportation** 運送過程中
- **arrive defective** 送達時有瑕疵
- **arrive damaged** 送達時有損毀

A Late Payment Apology Letter

寫作要點 Key Points:

Step 1: 對貨款滯納道歉。
Step 2: 簡短說明貨款滯納的具體原因。
Step 3: 承諾盡快支付貨款，並感謝對方諒解。

實際E-mail範例

| 寫信 ▼ | 刪除 | 回覆 ▼ | 寄件者： Ginny |

Dear Sir/Madam,

Introduction

With regard to the order for 200 sports jackets in June, I am sorry that we failed to make the payment on the scheduled date.

Body

The delay in processing the invoice and payment was due to the staff change in our finance department. The payment shall be completed by 5 pm today.

Closing

Please accept my sincere apology for delaying the payment.

Thanks for your understanding, and we look forward to working with you again.

Best regards,
Signature of Sender
Sender's Name Printed

E-mail中譯

寫信｜▼　刪除　回覆｜▼　寄件者：Ginny

您好：

開頭

有關我們於 6 月份向您購買 200 件運動外套的訂單，很抱歉沒有如期付款。

本體

由於本公司財務部門人事變動，延誤了收據和帳款的處理流程。這筆款項會在今天下午 5 點之前結清。

結尾

請接受我對貨款滯納的誠摯歉意。

感謝您的諒解，並期待繼續與您合作。

敬祝安康，
簽名檔／署名

各段落超實用句型

說明：畫底線部分的單字可按照個人情況自行替換

開頭 Introduction

❶ **We would like to apologize for delaying our payment.**
我們為貨款滯納向您致歉。

❷ **It is with regret that we were unable to make the payment on time.**
很抱歉我們無法準時支付帳款。

❸ **We apologize for not making the payment due last Friday.**
我們為滯納上週五到期的帳款道歉。

❹ **Please accept our apology for our tardiness in the payment.**
請接受我們對貨款滯納一事所表示的歉意。

⑤ We regret that you did not receive the payment expected.

很抱歉您沒有如期收到貨款。

⑥ We are terribly sorry for the delay in processing your payment.

對於延遲付款，我們非常抱歉。

本體 Body

❶ The delay of payment was resulted from the staff change in our finance department.

延遲付款乃本公司財務部門人事變動所致。

❷ Our finance department is very shorthanded and we're unable to process all the invoices in time.

我們的財務部門目前人力短缺，以至於我們無法及時處理所有的帳單。

❸ All invoices were mixed up due to the relocation of our office.

所有請款帳單因為辦公室搬遷全都混在一起了。

❹ We are trying our best to complete all our payments as soon as possible.

我們正盡全力儘快支付所有帳款。

❺ We assure you that we will transfer the payment in full no later than this Friday.

我們保證會在本週五之前將全額帳款轉給您。

❻ We had a bit of trouble with the service provider, but everything has been sorted out and you will receive your payment tomorrow at the latest.

我們在服務提供者那邊遇到了一點問題，但現在一切都處理就緒了。您最晚明天就會收到款項。

結尾 Closing

❶ Sorry for the trouble and inconvenience.

造成困擾及不便，很抱歉。

❷ Sorry for our tardiness in the payment.

貨款滯納，很抱歉。

❸ We apologize again for our late payment.

再次為貨款滯納向您致歉。

❹ We hope this incident won't jeopardize our future dealings.

希望這次事件不會影響我們未來的合作。

❺ Please contact me immediately if you do not receive the payment by <u>tomorrow</u>.

若您<u>明天</u>還未收到帳款，請立刻與我聯繫。

❻ Again, we apologize for the inconvenience this may have caused you.

對於可能造成您的不便，我們再度致上歉意。

 # 不可不知的實用E-mail字彙

- **tardiness in the payment** 貨款滯納
- **late payment** 延遲繳款
- **delay the payment** 延遲繳款
- **overdue payment** 過期未繳的帳款
- **make the payment on time** 準時支付帳款

05 貨物錯誤道歉

An Apology Letter for Shipping Mistakes

寫作要點 Key Points:

Step 1: 為貨物錯誤道歉。
Step 2: 提供具體解決方式。
Step 3: 懇請對方諒解，並希望仍有合作機會。

 實際E-mail範例

| 寫信 ▼ | 刪除 | 回覆 ▼ | 寄件者： Ginny |

Dear Sir/Madam,

Introduction

We are writing this letter to apologize for delivering you the wrong sized dress.

Body

We will certainly be responsible for the error. Our fast courier will send you the correct commodity within 24 hours and retrieve the incorrect item at the same time.

To make up for our mistake, we are also sending you a 30% off coupon for your use on your next purchase.

Closing

We are sorry for causing you inconvenience.

Please do continue purchasing products from us.

Thank you very much.

Best regards,
Signature of Sender
Sender's Name Printed

E-mail中譯　[寫信 ▼]　[刪除]　[回覆 ▼]　寄件者：[Ginny]

您好：

開頭

我們寫這封信是要向您致歉，因為我們寄給您尺寸錯誤的洋裝。

本體

我們當然會對這次錯誤負責。我們的快遞會在 24 小時之內將正確商品遞送給您，並同時取回錯誤商品。為了彌補我們的失誤，我們也會送給您一張七折優惠券，供您下次購物使用。

結尾

造成您的不便，我們很抱歉。

請務必繼續購買我們的商品。

非常感謝您。

敬祝安康，
簽名檔／署名

 # 各段落超實用句型

說明：畫底線部分的單字可按照個人情況自行替換

開頭 Introduction

❶ **We regret that we have made a mistake in your shipment.**
很抱歉我們在為您出貨時出了點錯。

❷ **We are very sorry for delivering your order to the wrong address.**
將您訂購的物品送到錯誤的地址，我們感到非常抱歉。

❸ **We sincerely apologize for sending you a wrong commodity.**
我們誠心為送錯您的商品致上歉意。

❹ **We are writing this letter to apologize for a shipping error.**
我們寫這封信是為了出貨問題而向您致歉。

❺ We would like to apologize for shipping you the incorrect quantity.

我們想要為出貨數量錯誤向您道歉。

❻ We noticed that we have delivered the wrong sized shirt to you.

我們發現我們寄了尺寸錯誤的襯衫給您。

❼ We are truly sorry for the mix-up in your orders.

對於搞混您的訂單，我們非常抱歉。

本體 Body

❶ We will send you the rest of your commodities today.

我們將會在今天為您寄出您訂購的其他商品。

❷ We will be immediately sending you the correct commodities.

我們將立刻寄送正確的商品給您。

❸ We will send your order to the correct address by fast courier within the next 24 hours.

我們將會請快遞在二十四小時之內將您訂購的商品送到正確的地址。

❹ Our courier will send you the correct item and retrieve the incorrect item within the next two days.

本公司快遞人員會在兩天內為您送上正確的商品，並取回錯誤的商品。

❺ The correct items have been shipped and they shall arrive no later than this Wednesday.

正確的商品已經寄出，應該在本週三之前就會到。

❻ Please expect the correct order to arrive at your doorstep early tomorrow.

正確的訂單將在明天一早送到您那裡。

結尾 Closing

❶ We are sorry for causing you trouble and inconvenience.

為您帶來麻煩與不便，我們很抱歉。

❷ We apologize again for the unintended error.

再次為此無心之過向您致歉。

❸ To make up for our mistake, we are giving you <u>20%</u> off on your next purchase.
為了彌補我們的失誤，我們將提供您下次購物<u>八折</u>優惠。

❹ We hope you continue doing business with us.
希望您繼續與我們合作。

❺ We apologize for our mistake and hope that this does not affect our business partnership in any way.
我們很抱歉犯了這個錯誤，希望這不會影響到我們事業上的合作。

 # 不可不知的實用E-mail字彙

- **shipping error** 出貨錯誤
- **incorrect quantity** 數量錯誤
- **incorrect item** 錯誤商品
- **correct item** 正確商品
- **wrong address** 錯誤地址
- **correct address** 正確地址

06 意外違反合約道歉

An Apology Letter for Unintended Breach of Contract

 實際E-mail範例

寫信 ▼	刪除	回覆 ▼	寄件者： Ginny

Dear Sir/Madam,

Introduction

We are writing this letter to express our sincere apology for accidentally breaching our contract with you.

Body

We took some time to find out what happened, and hereby would like to apologize for revising your article without your consent.

Closing

We assure you that this kind of behavior will never occur again.

We sincerely hope you would forgive us for our unintentional error.

Best regards,
Signature of Sender
Sender's Name Printed

E-mail中譯　　寫信 ▼　　刪除　　回覆 ▼　　寄件者： Ginny

您好：

開頭

我們寫這封信的目的是要對意外違反合約一事向您致上誠摯的歉意。

本體

我們花了點時間找出發生什麼事，並希望以此方式，對未經您同意就擅自修改您的文章，向您道歉。

結尾

我們向您保證這樣的行為將不會再次發生。

誠摯希望您能原諒我們此次無心的過失。

敬祝安康，
簽名檔／署名

各段落超實用句型

說明：畫底線部分的單字可按照個人情況自行替換

開頭 Introduction

❶ **We apologize for our unintended error.**
我們為我們的無心之過向您致歉。

❷ **Please understand that we did not breach our contract on purpose.**
請明白我們並非蓄意違反合約。

❸ **We are sorry for unintentionally breaching our contract with you.**
我們很抱歉，不小心違反了與您的合約。

❹ **We would like to express our regret for committing unintentional behavior that breached our contract with you.**
我們想表達我們對於無意中違反合約的行為之懊悔與歉意。

❺ With this letter we would like to request for your forgiveness if we have unintentionally breached our contract with you.

如果我們不小心違反了與您的合約，我們希望能藉由這封信請求您的原諒。

❻ We are incredibly sorry about the unintentional breach in our contract with you.

意外違反了與您的合約，我們真的非常抱歉。

本體 Body

❶ We are regretful for reprinting your book without your consent.

我們很抱歉未經您的允許就再版您的書。

❷ We shouldn't have discussed the project with a third party.

我們不應該與第三方討論專案內容。

❸ We are ashamed of failing to protect your personal information.

未盡保護您個人資料之責，我們感到萬分愧疚。

❹ We take the blame for revising your article without your consent.

擅自修改您的文章，是我們的錯。

❺ We took swift action to prevent any further unauthorized access to customers' personal information as soon as we could.

我們已經盡快採取行動，預防任何未經認證取得顧客個人資料之途徑。

❻ This is a truly unprofessional error and we are sincerely sorry about the inconvenience this has caused you.

這是個很不專業的錯誤，造成了您的不便，我們非常抱歉。

結尾 Closing

❶ Please forgive us for our unintended error.

請原諒我們無心的過失。

❷ We are sorry that this incident occurred and caused you so much trouble.

很抱歉發生這種事，造成你們這麼多困擾。

❸ **We assure you that this kind of behavior will never occur again.**
我們向您保證這樣的行為將不會再次發生。

❹ **We will take every effort to prevent such an error from occurring in the future.**
我們會盡一切努力避免未來再發生此錯誤。

❺ **We would like to request a meeting with you to discuss this breach of contract.**
我們希望能跟您開個會討論此次違反合約一事。

❻ **If you would like, we can hold a meeting to discuss in detail ways we can remedy the situation.**
若您願意,我們可以開會詳細討論補救的方案。

 # 不可不知的實用E-mail字彙

- **breach of contract** 違反合約
- **breach the contract** 違反合約
- **unintended error** 無心的過失
- **on purpose** 故意地
- **request for forgiveness** 請求原諒
- **take the blame for sth.** 為某事承擔責任

07 感謝活動款待

A Hospitality Thank-you Letter

寫作要點 Key Points:

Step 1: 說明來信目的與某個活動有關。
Step 2: 感謝對方在活動中的款待。
Step 3: 表示希望有機會可以回報對方的款待。

實際E-mail範例

寫信 ▼	刪除	回覆 ▼	寄件者： Ginny

Dear Sir/Madam,

Introduction

I am writing this letter in regards to your new product launch party on May 15th, 2017. Thank you for having me as your VIP guest.

Body

I appreciate your kind hospitality at the party. I had a wonderful time and I am grateful for having a chance to get to know some of your honorable guests.

Closing

Thanks again for your warm hospitality.

I look forward to the opportunity to reciprocate in the near future.

Best regards,
Signature of Sender
Sender's Name Printed

E-mail中譯

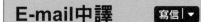

寄件者：**Ginny**

您好：

開頭

我是為了您在 2017 年 5 月 15 日舉行的新品上市派對而寫信給您。感謝您邀請我以貴賓身份出席。

本體

感謝您在派對上的好意款待。我度過了愉快的時光，並且很感謝能有機會結識幾位您的貴賓。

結尾

再次感謝您的溫馨款待。

期待能很快有回報您款待的機會。

敬祝安康，
簽名檔／署名

各段落超實用句型

說明：畫底線部分的單字可按照個人情況自行替換

開頭 Introduction

❶ Thank you for having me as your VIP guest.

感謝您邀請我以貴賓身份出席。

❷ Thank you for all the kind hospitality that you showed me at your annual banquet.

感謝您在您的尾牙餐會上對我的款待。

❸ I am writing this letter to express my appreciation for your warm hospitality and friendliness.

寫這封信是為了表達我對您溫暖的款待及友誼的感激之情。

❹ We would like to express our sincere thanks to you for having us at your charity dinner party.

我們想表達我們對您邀請我們到您的慈善晚宴的衷心感謝。

❺ I am writing to thank you for all your attendance during my trip to Tokyo.

寫這封信是要為了要感謝您在我到東京出差時對我的照顧。

❻ We wanted to thank you for inviting us to your opening ceremony.

我們想感謝您邀請我們參加貴公司的開幕典禮。

本體 Body

❶ I appreciate your kind hospitality at the party.

感謝您在派對上對我的好意款待。

❷ We had a great time at your annual banquet.

我們在貴公司的尾牙餐會度過了愉快的時光。

❸ Your warm hospitality made my business trip to New York a wonderful one.

您溫暖的款待使我在紐約有了一次非常美好的出差經驗。

❹ Thank you for taking the time to show me around your headquarters.

感謝您花時間帶我參觀貴公司的總部。

❺ We felt at home throughout the entire party.

我們整場宴會都感覺非常自在。

❻ It was a truly wonderful party and I enjoyed myself thoroughly.

這場派對非常棒,我玩得很開心。

結尾 Closing

❶ Thanks again for your warm hospitality.

感謝您的溫馨款待。

❷ We hope to reciprocate in the near future.

希望很快能回報您的款待。

❸ **Please accept my heartfelt thanks to you for making my trip an enjoyable one.**

請接受我發自內心的感謝您使我這趟出差非常愉快。

❹ **We hope you can visit us sometime soon and let us reciprocate your hospitality.**

我們希望您能找時間很快來拜訪我們，讓我們回報您的款待。

❺ **Perhaps, if you would like, I could return the favor by showing you around our headquarters as well.**

如果您願意，或許我可以帶您參觀我們的公司總部作為報答。

 不可不知的實用E-mail字彙

• **warm hospitality**　溫暖的款待
• **have a wonderful time**　度過美好的時光
• **reciprocate sb.'s hospitality**　回報某人的款待
• **attendance**　伺候；照料
• **heartfelt thanks**　發自內心的感謝

08 感謝下訂／合作

A Business Thank You Letter

寫作要點 Key Points:

Step 1: 感謝對方的訂單／合作。
Step 2: 承諾完成訂單，或確認訂單內容。
Step 3: 表示希望未來能繼續合作。

實際E-mail範例

| 寫信 ▼ | 刪除 | 回覆 ▼ | 寄件者： Ginny |

Dear Sir/Madam,

Introduction

We are pleased that you have chosen JBS for your bathroom equipment purchase. JBS is delighted to welcome you as our new client.

Body

We hope you are satisfied with the excellent quality of your new bathtub.

Please be kindly advised that all our products have two year's guarantee from the date of purchase.

Closing

Should you have any questions concerning our products, please call 1234-5678 for our customer service, which is available 24/7.

Thank you again for your purchase.

We look forward to serving you again.

Best regards,
Signature of Sender
Sender's Name Printed

E-mail中譯

寫信▼　刪除　回覆▼　寄件者：Ginny

您好：

開頭

很高興您決定選購 JBS 的衛浴設備產品，我們很開心歡迎您成為我們的新客戶。

本體

我們希望您很滿意您新浴缸的優良品質。

在此貼心提醒您，我們所有的產品都包含自購買日起 24 個月的保固期。

結尾

萬一您對本公司產品有任何問題，請撥打本公司 24 小時全年無休的客服專線 1234-5678。

再次感謝您的購買。

期待能再次為您服務。

敬祝安康，
簽名檔／署名

各段落超實用句型

說明：畫底線部分的單字可按照個人情況自行替換

開頭 Introduction

❶ Thank you very much for ordering <u>clothes and accessories</u> from us.

非常感謝您向我們訂購<u>服飾與配件</u>。

❷ We are pleased that you have chosen our products for your <u>office equipment</u> purchase.

很高興您選擇本公司產品作為您的<u>辦公室設備</u>。

❸ We are very delighted to welcome you as our new client.

我們很高興歡迎您成為本公司的新客戶。

❹ We have received your order. Thank you for choosing our products.
我們已經收到您的訂單了。感謝您選擇我們的產品。

❺ Thanks for ordering our products from our website.
感謝您在本公司網站上訂購產品。

本體 Body

❶ We hope you are happy with the excellent quality of your <u>new washing machine</u>.
我們希望您對您選購的<u>新洗衣機</u>的優良品質感到很滿意。

❷ We hope you have as much fun using it as we did designing it!
我們希望您在使用這項產品時，就如同我們在設計它時一樣開心。

❸ Please be kindly advised that all our products have <u>two year's</u> guarantee from the date of purchase.
在此貼心通知您，我們所有的產品都有自購買日起 <u>24 個月</u>的保固期。

❹ Please note that your order will be sent to you <u>COD</u> within <u>five working days</u>.
在此通知您，您訂購的商品會在<u>五個工作天內</u>以<u>貨到付款</u>的方式寄送給您。

❺ We are pleased to inform you that your order has been processed and will be delivered to you upon receipt of your payment.
很高興通知您，我們已經受理您的訂單，並且在收到您的訂單款項後就會立刻寄送給您。

結尾 Closing

❶ We appreciate the opportunity to do business with you.
我們感謝有與您合作的機會。

❷ We look forward to a long and mutually beneficial relationship with you.
我們期待與貴公司建立長久的互惠合作關係。

❸ If you are interested in any other commodities, please let me know.
如果您對本公司其他產品有興趣，請與我聯絡。

❹ Should you have any questions concerning our products, please contact our 24/7 Customer Service.

萬一您對本公司產品有任何問題，請撥打本公司 24 小時全年無休的客服部聯絡專線。

❺ Let me know if you're interested in any other commodities.

如果您對任何其他產品有興趣，請與我聯絡。

❻ We are grateful for your patronage and hope you will continue to do business with us in the future.

我們很感謝您的惠顧，希望您未來也會持續與我們合作。

 ## 不可不知的實用E-mail字彙

- **12-month guarantee** 十二個月保固期
- **24-month warranty** 兩年品質保證
- **customer service** 客服部
- **24/7 (twenty-four seven)** 24 小時全年無休
- **COD= cash on delivery** 貨到付款
- **long and mutually beneficial relationship** 長久的互惠合作關係

A Letter of Appreciation for Assistance at Work

寫作要點 Key Points:

Step 1: 感謝對方提供業務上的協助。
Step 2: 說明對方的協助，對完成哪些事項有所幫助。
Step 3: 表示希望有回報對方的機會。

 實際E-mail範例

寫信 ▼	刪除	回覆 ▼	寄件者： Ginny

Dear Sir/Madam,

Introduction

I, with this letter, would like to express my gratitude to you for providing us your technical assistance in this promotion project.

Body

If it weren't for your support and assistance, this promotion project wouldn't have been so successful.

Closing

Please do let us know if there's anything we can do for you in return for all your help.

Thank you again, and wish you all the best.

Best regards,
Signature of Sender
Sender's Name Printed

E-mail中譯　寫信▼　刪除　回覆▼　寄件者：Ginny

您好：

開頭

我想藉這封信向您表達，對於您在本次促銷專案上的技術支援，我十分感激。

本體

如果沒有您的支持與協助，這次的促銷專案不會如此成功。

結尾

如果有任何我們能回報您的事，請儘管讓我們知道。

再次感謝您，並祝您一切順心如意。

敬祝安康，
簽名檔／署名

 # 各段落超實用句型

說明：畫底線部分的單字可按照個人情況自行替換

開頭 Introduction

❶ I would like to express my gratitude to you for providing us your technical assistance in this promotion project.

對於您在這次促銷專案上提供的技術支援，我想向您致上誠摯的謝意。

❷ We appreciate your assistance in preparing this sales presentation.

我們想感謝您在準備這次業務簡報上的協助。

❸ We are very thankful for your assistance in sending the product sample to us immediately upon request.

感謝您一經我們要求就立刻寄送樣品給我們。

❹ I would like to than you for covering my shift yesterday.

我想謝謝你昨天幫我代班。

❺ Please accept our sincere thankfulness for your help and support on the project presentation.

請接受我們對您在專案簡報上給予的協助與支持，致上誠摯的謝意。

❻ I am writing to thank you for the constant guidance you have provided me in the past month.

我寫這封信，是要感謝您在過去這個月來給我的持續指導。

本體 Body

❶ The project wouldn't have gone so well if it weren't for your guidance and instruction.

若非您的引導與指點，這次專案不會進行得如此順利。

❷ Your technical assistance helped us run a successful business event.

您的技術支援協助我們舉辦了一場成功的商業活動。

❸ Your assistance made the event a successful one.

您的協助讓這次活動非常成功。

❹ I appreciate your willingness to help outside your current position.

感激您願意在現任職務之外提供幫助。

❺ Thank you for sharing your experience and helping me create a successful and professional presentation.

感謝您分享您的經驗，並幫助我完成了一次成功且專業的發表會。

❻ If it weren't for your dedication, we would not have been able to complete this ambitious project in such a short time.

若沒有您的投入貢獻，我們無法在這麼短的時間內完成這麼大的計畫。

結尾 Closing

❶ Please let me know whenever you need anything from me.

有什麼事情需要幫忙，請一定告訴我。

❷ Let me know if there is anything I can do to return your favor.

有什麼我可以回報你的事，請告訴我。

❸ **I look forward to any opportunities to return the favor.**
我期待任何能回報你恩情的機會。

❹ **Please do not hesitate to let me know if I can be of any assistance to you.**
如果有什麼我可以幫得上忙的，請儘管告訴我。

❺ **Thank you again for your time and contribution on this** <u>**project**</u>**.**
再次感謝您為這次<u>專案</u>付出的時間和貢獻。

❻ **I will be glad to return the favor in the future, so if you need anything from me, do not hesitate to ask.**
我很樂意在未來回報您，若您需要我的幫助，請不吝告知我。

 # 不可不知的實用E-mail字彙

- **technical assistance** 技術支援
- **guidance and instruction** 引導與指點
- **support and assistance** 支持與援助
- **cover sb.'s shift** 幫某人代班
- **return the favor** 回報恩情

Part 10

誠心祝賀篇

01 祝賀職位升遷

A Job Promotion Congratulation Letter

寫作要點 Key Points:

Step 1: 恭賀對方獲得陞遷。
Step 2: 讚揚對方工作表現終於實至名歸。
Step 3: 期許對方未來工作表現，並祝福對方事業成功。

實際E-mail範例

寫信 ▼	刪除	回覆 ▼	寄件者： Ginny

Dear Sir/Madam,

Introduction

I am so happy to hear about your promotion to the Chief Editor position. Please accept my congratulations on your promotion!

Body

I have always been impressed by your outstanding performance at work and your brilliant leadership. I must say that I was not surprised at your promotion at all because this recognition of your ability is well deserved.

Closing

I am already looking forward to hearing about your promotion celebration party. Congratulations again, and wish you continued success in your new position.

Best regards,
Signature of Sender
Sender's Name Printed

E-mail中譯

寫信 ▼　刪除　回覆 ▼　寄件者：Ginny

您好：

開頭

得知您被晉升為總編輯，我實在非常開心！

請接受我對您晉升的祝賀之意！

本體

我一直都很欣賞您在工作上的傑出表現以及您優異的領導才能。我必須説我對您的升遷一點都不感到意外，因為您的工作能力值得肯定。

結尾

我已經開始期待您的晉升慶祝派對了。

再次恭喜您，並祝您在新職位上大展宏圖。

敬祝安康，
簽名檔／署名

各段落超實用句型

説明：畫底線部分的單字可按照個人情況自行替換

開頭 Introduction

❶ **We are glad to know that you have been promoted to the Manager of the Research and Development Department.**

得知您已經被晉升為研發部經理，我們感到非常高興。

❷ **It's a thrill to hear that you have been appointed the Manager of the Marketing Department.**

聽到您已經被任命為行銷部經理，真是令人感到興奮。

❸ **Please accept my heartiest congratulations on your promotion as the District Manager.**

請接受我誠心恭賀您榮升區經理。

❹ I am writing this letter to congratulate on your promotion.
我寫這封信是要恭賀你升職。

❺ Big congratulations on your recent promotion!
大大地恭喜您近日榮獲升遷！

本體 Body

❶ Getting a promotion is definitely a big career milestone.
獲得晉升絕對是事業上一個重要的里程碑。

❷ Your dedication at work definitely deserves this recognition.
你對工作的付出絕對值得這樣的肯定。

❸ I have always been impressed by the high quality of your work, so I wasn't surprised at your promotion at all.
我一直很欣賞你優良的工作品質，所以我對你的晉升一點都不感到意外。

❹ I am happy for your achievement and you truly deserve the promotion.
我為您的成就感到高興，而此次的晉升，您確實當之無愧。

❺ Your outstanding performance at work is impressive. This recognition of your ability is well deserved.
你傑出的工作表現讓人激賞，你的工作能力值得這項肯定。

❻ I'm delighted to hear that you have received the promotion you deserve and I'm positive that you will play your new role perfectly.
我很高興聽說您獲得升遷，這是您應得的，我相信您在新的職位肯定會做得很好。

結尾 Closing

❶ The promotion deserves a fantastic person like you.
這個晉升機會理當給予一個像您這樣傑出的人。

❷ It was the promotion that you rightfully deserved.
這次晉升對你而言是理所當然的。

❸ I look forward to being invited to your promotion celebration party.
我期待能受邀參加您的晉升慶祝派對。

❹ **Wish you continued success in your new position.**
祝您在新職位上大展宏圖。

❺ **I am confident that your department will continue growing under your leadership.**
我相信貴部門在您的帶領下會繼續成長茁壯。

❻ **I wish you continued success in your future endeavors and a lucrative career ahead of you.**
祝福您未來的事業成功，職涯大有斬獲。

 # 不可不知的實用E-mail字彙

- **promotion** 晉升
- **promotion celebration party** 晉升慶祝派對
- **career milestone** 事業上的里程碑
- **recognition** 肯定
- **outstanding performance at work** 工作上的傑出表現
- **high quality of work** 優良的工作品質
- **brilliant leadership** 優異的領導能力
- **hard work** 認真工作
- **dedication at work** 對工作的投入
- **well-deserved** 應得的；理所當然的

An Office Relocation Congratulation Letter

寫作要點 Key Points:

Step 1: 恭賀對方公司喬遷。
Step 2: 表達對新辦公室的期許。
Step 3: 祝賀對方在新辦公地點宏圖大展。

 實際E-mail範例

寫信 ▼	刪除	回覆 ▼	寄件者： Ginny

Dear Sir/Madam,

Introduction

On behalf of JBS Company, I am sending my heartiest congratulations to you on your move to the new office building.

Body

We are happy to know that your new office is a more spacious and greater place to work in. I am sure that your employees will perform their tasks even better in the new office.

Closing

We look forward to having opportunities to visit your new office.

Wish you continued success in the new office.

Best regards,
Signature of Sender
Sender's Name Printed

E-mail中譯

寫信 ▼　刪除　回覆 ▼　寄件者：Ginny

您好：

開頭

謹代表 JBS 公司，我要對貴公司喬遷至新辦公大樓，向您獻上我最誠心的恭賀。

本體

很高興得知您的新辦公室是個更寬敞、更棒的工作地點。我相信您的員工在新辦公室一定會有更優異的工作表現。

結尾

我們期待能有機會到您的新辦公室拜訪您。

祝您在新辦公室繼續大展宏圖。

敬祝安康，
簽名檔／署名

各段落超實用句型

說明：畫底線部分的單字可按照個人情況自行替換

開頭 Introduction

❶ **I am writing this letter on behalf of <u>JBS Company</u> to congratulate you on your move to the new office.**
謹代表 JBS 公司，寫此信恭喜貴公司喬遷新址。

❷ **Please accept our sincere congratulations on your new office space.**
請接受我們對您擁有新辦公地點的誠摯恭賀之意。

❸ **I am glad to hear that you have finally moved to a new office.**
很高興得知貴公司終於喬遷至新辦公室的消息。

❹ Congratulations on moving to the new office!

恭喜貴公司喬遷新址！

❺ I am pleased to hear that you have moved to your new office building.

很高興知道貴公司已經遷至新辦公大樓。

本體 Body

❶ I heard your new office is a more spacious and greater place to work in.

聽說您的新辦公室是個更寬敞、更棒的工作地點。

❷ I am sure that your employees will perform even better in the new office.

我相信您的員工在新辦公室一定會有更優異的工作表現。

❸ I am confident that your business will be even more prosperous in your new office.

我相信您的事業在新辦公室一定會更加成功。

❹ Your new office space will definitely open new doors for your business.

您的新辦公室一定能為您的事業開創新局。

❺ I believe moving into the new office building will boost your business.

我相信遷到新辦公大樓，一定會讓貴公司生意興隆。

❻ I'm sure that your new office, with its excellent location and ample space, will bring you even more business opportunities.

您的新辦公地點擁有極佳的位置與充足的空間，相信將能為您帶來更多事業良機。

結尾 Closing

❶ We wish you best luck and success in your new place of work.

我們祝您在新辦公室順心如意，事業成功。

❷ **Wish you continued success in your new office.**
祝您在新辦公室越來越成功。

❸ **Congratulations again on your new office and wish you the best in the future.**
再次恭喜您遷至新辦公室,並祝您未來一切順利。

❹ **Hope your new working place will bring you prosperity and success in the long run.**
希望您的新辦公室會讓您生意興隆,事業成功。

❺ **We look forward to visiting you in your new office soon.**
期待很快可以到您的新辦公室拜訪您。

❻ **I look forward to hearing about your booming business in the new location.**
我很期待聽到您公司在新地點蓬勃發展。

 # 不可不知的實用E-mail字彙

- **new office** 新辦公室
- **new working place** 新辦公地點
- **new factory** 新工廠
- **new location** 新地點
- **new manufacturing plant** 新製造工廠
- **better place to work on** 更棒的工作場所

A Holiday Greeting Letter

寫作要點 Key Points:

Step 1: 表達來信目的為給予佳節祝賀。
Step 2: 感謝對方一直以來的合作。
Step 3: 祝福對方佳節愉快。

 實際E-mail範例

寫信 ▼　刪除　回覆 ▼　　寄件者： Ginny

Dear Sir/Madam,

Introduction

On behalf of JBS Company, I am writing to send my warmest greetings for Chinese New Year to you and your family.

Body

Thank you for giving us the opportunity to work with you. We appreciate your assistance and support in this past year.

Closing

We wish you the best of holidays and we wish you health, happiness and much success in the coming year.

Best regards,
Signature of Sender
Sender's Name Printed

E-mail中譯　寫信▼　刪除　回覆▼　寄件者：Ginny

您好：

開頭
謹代表 JBS 公司，以本信向您與您的家人致上最溫暖的農曆春節問候。

本體
感謝您給我們與您合作的機會。我們很感謝貴公司在過去這一年對我們的協助與支持。

結尾
祝您有個最棒的假期，並且預祝您在未來的一年健康、快樂、諸事成功。

敬祝安康，
簽名檔／署名

 # 各段落超實用句型

說明：畫底線部分的單字可按照個人情況自行替換

開頭 Introduction

❶ **On behalf of JBS Company, I am sending my warmest greetings for the Mid-Autumn Festival.**
我謹代表 JBS 公司向您致上最溫暖的中秋節問候。

❷ **With this letter I would like to convey warm greetings for the coming holidays.**
藉由此信，我想為了即將到來的佳節獻上溫暖的問候。

❸ **I am writing to send my holiday greetings to you and your staff.**
此信的撰寫目的是為了向您及您的員工獻上我的佳節問候。

❹ Happy Chinese New Year! I hope you have a holiday that fills your heart with joy.

春節快樂！希望您有個充滿喜樂的佳節。

❺ Please accept my heartiest wishes for the coming holidays.

請接受我為即將到來的假期所獻上的誠心祝福。

本體 Body

❶ Thank you for giving us the opportunity to work with you this year.

感謝您這一年給我們與您合作的機會。

❷ Thank you for your business this year.

感謝您這一年的合作。

❸ It has been a pleasure working with you this year.

很高興能在這一年與您合作。

❹ We treasure our relationship with you and appreciate your business in this past year.

我們很珍惜與您的合作關係，並感謝過去一年來您的惠顧。

❺ We value our friendship and thank you for your trust and collaboration.

我們很珍惜與您的友誼，並感謝您的信任與合作。

❻ We appreciate your patronage this year and look forward to your continued patronage in the coming year.

感謝您今年的惠顧，並期待您來年繼續惠顧。

❼ It has been a wonderful year working with you, and we look forward to another year of close partnership.

與您共事的這一年非常開心，我們很期待明年也繼續與您緊密合作。

結尾 Closing

❶ Best wishes from all the JBS staff.

JBS 全體員工在此獻上最佳祝福。

❷ **Please accept our warmest wishes for a wonderful holiday.**
請接受我們在美好佳節時對您獻上最誠摯的祝福。

❸ **Wish you health and happiness in your new year.**
祝您在新的一年裡健康快樂。

❹ **I wish you and all your employees a pleasant holiday season.**
祝您與您所有同仁有個愉快的佳節。

❺ **We wish you the best of holidays and much success in the coming year.**
祝您有個最棒的假期，並祝您來年諸事成功。

❻ **We wish you and your excellent staff a very merry holiday season.**
祝福您與您優秀的員工佳節愉快。

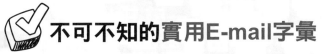 # 不可不知的實用E-mail字彙

- **Thanksgiving** 感恩節
- **Christmas** 聖誕節
- **New Year** 元旦
- **Chinese New Year** 中國新年
- **Dragon Boat Festival** 端午節
- **Mid-Autumn Festival** 中秋節
- **holiday greetings** 佳節問候

04 祝賀開業、公司擴展

A New Business Congratulation Letter

寫作要點 Key Points:

Step 1: 恭賀對方開業或公司擴展。
Step 2: 讚揚對方能力。
Step 3: 祝福對方大展宏圖，並表明期待有機會能合作。

 實際E-mail範例

寫信 ▼	刪除	回覆 ▼	寄件者：	Ginny

Dear Sir/Madam,

Introduction

It's a thrill to hear that you are opening a new branch store in Shanghai.

Please accept my heartiest congratulations on your business expansion to China.

Body

It's great to see that you are accomplishing your goal of business expansion so fast.

Closing

Congratulations again.

Good luck and best wishes on your business success.

Best regards,
Signature of Sender
Sender's Name Printed

E-mail中譯

寫信▾　刪除　回覆▾　寄件者：Ginny

您好：

開頭

聽到您要在上海開設新分店，真是令人興奮。

請接受我對您事業拓展至中國，致上最真誠的祝賀！

本體

看到您那麼快就達成了拓展事業版圖的目標，真是太棒了。

結尾

再次恭喜您！

祝您一切順利，並為您的事業成功獻上最佳的祝福。

敬祝安康，

簽名檔／署名

 各段落超實用句型

　　説明：畫底線部分的單字可按照個人情況自行替換

開頭 Introduction

❶ **I am glad to hear that you are starting your own business.**

很高興得知您已經創業了。

❷ **Congratulations on the opening of your own business!**

恭喜您創業！

❸ **Congratulations on the recent opening of the new branch of your store in Kaohsiung.**

恭喜您近日在高雄的分店開張了。

❹ **Please accept my heartiest congratulations on your new business.**

請接受我對您的新事業的最誠心恭賀。

❺ I am writing to convey my sincere congratulations on your new venture.

我是寫信來表達對您新事業的誠摯祝賀。

❻ This is to congratulate you on your brand-new start in business ventures.

這封信是要恭賀您在事業上的全新開始。

本體 Body

❶ Opening your own business is absolutely an extraordinary milestone.

開創自己的事業絕對是個了不起的里程碑。

❷ I am so proud to have been your colleague at JBS Company.

作為你在 JBS 公司的同事，我深感驕傲。

❸ It's great to see that you are accomplishing goals of success so fast.

很高興看到您如此快速就達成了成功的目標。

❹ I have no doubt that your new venture will be successful under your leadership.

我堅信您的新事業會在您的帶領之下成功茁壯。

❺ It has long been your dream to start your own business. Congratulations on making your dream a reality.

創業一直是你長久以來的夢想。恭喜你美夢成真！

❻ I know that this has always been what you wanted, and it's simply delightful to see it finally come true.

我知道這是您一直以來的夢想，終於成真了，讓我非常開心。

結尾 Closing

❶ Wish you all the success towards your goal.

祝福您成功地往您的目標邁進！

❷ I look forward to doing business with you in the near future.

期待不久的將來能與您合作。

❸ **Good luck with your new business.**
祝福您的新事業順利。

❹ **Wishing you the best for your startup.**
在此對您創業致上最佳祝福。

❺ **I wish you much luck on your new business venture.**
祝福您新事業一切順利。

❻ **I look forward to all the future business collaboration opportunities this will bring.**
期待未來你我將因此有許多生意上合作的機會。

 # 不可不知的實用E-mail字彙

* **start one's own business** 自己創業
* **new business** 新事業
* **new venture** 新事業
* **new branch office** 新分公司
* **new branch restaurant** 新餐廳分店
* **new branch store** 新分店

Part11

其他重要信件

01 請求聯絡與回覆

A Response Request Letter

 實際E-mail範例

| 寫信 ▼ | 刪除 | 回覆 ▼ | 寄件者： Ginny |

Dear Sir/Madam,

Introduction

I am George Lin, the new sales representative of JBS Company.

I missed a call from your office this morning while I was in a meeting with one of my clients.

Body

As I am not aware of your office extension number, or your other contact information, I am writing this letter to you to request your response.

You can contact me by calling my office number 2123-4567 at extension 123. I can also be reached by my email, georgelin@jbs.com or by my cellphone 0912345678 during the office hours.

Closing

Look forward to hearing from you soon.

Best regards,
Signature of Sender
Sender's Name Printed

E-mail中譯 寫信▼ 刪除 回覆▼ 寄件者：Ginny

您好：

....

開頭

我是 JBS 公司的新業務代表林喬治。

由於正在跟客戶開會的緣故，今天上午我沒有接到您辦公室打來的電話。

本體

由於我不知道您的公司分機號碼，或您其他的聯絡資訊，因此寫這封信以請求您的回覆。

您可以撥打公司電話 2123-4567 轉分機 123 與我聯絡，也可以在上班時間以電子郵件 georgelin@jbs.com 或手機 0912345678 與我聯繫。

結尾

期待能很快得到您的回覆。

....

敬祝安康，
簽名檔／署名

 各段落超實用句型

說明：畫底線部分的單字可按照個人情況自行替換

開頭 Introduction

❶ **I missed a call from your office this morning.**
我沒接到今天上午您的辦公室打來的電話。

❷ **I've got a few missed calls from your cellphone number.**
我接到來自您手機號碼的數通未接來電。

❸ **I have been trying to get in touch with you for the past few days.**
過去幾天我一直試著想要聯絡您。

❹ **I called your office but failed to reach you.**
我打電話到您辦公室，但沒有聯絡上您。

❺ I am sending you this email because I tried to call you this morning but my call was left unanswered.

我寫這封信給您，是因為我今天早上試著打電話給您，但是卻沒人接。

❻ I have sent you several emails during the past week, but have never received any reply.

過去這一週，我寄了好幾封電子郵件給您，但完全沒有收到回覆。

本體 Body

❶ Please contact me at your earliest convenience.

麻煩您盡快與我聯絡。

❷ Please return my call as soon as possible because it's kind of an emergency.

因為有點緊急，請您立刻回我電話。

❸ Please return my call or email as soon as you are available.

請您一有時間就回我電話或是電子郵件。

❹ There's something very important that I need to discuss with you immediately.

我有非常重要的事情要立刻與您討論。

❺ I can be reached by my cellphone at any time.

任何時候，你都可以打我手機找到我。

❻ There is an urgent matter I need to discuss with you, and I would appreciate it if you could contact me as soon as possible.

我有一件急事必須與您討論，所以請盡快與我聯絡，我將感激不盡。

結尾 Closing

❶ Thank you for your attention.

感謝您的留意。

❷ Look forward to hearing from you soon.

期待能盡快得到您的回覆。

❸ Look forward to receiving your call.

期待接到您的來電。

❹ **Look forward to your email.**
期待收到您的來信。

❺ **Look forward to your timely reply.**
期待您的立即回覆。

❻ **I will be looking forward to talking to you very soon.**
我很期待不久後能夠與您討論。

 不可不知的實用E-mail字彙

• **miss a call** 未接一通電話
• **missed call** 未接來電
• **call left unanswered** 電話未接聽
• **get in touch with sb.** 與某人取得聯繫
• **return sb.'s call** 回某人電話
• **return sb.'s email** 回某人信件

02 向主管或人事單位請假

A Leave Request Letter

寫作要點 Key Points:

Step 1: 提出請假要求。
Step 2: 說明請假原因，與請假期間的工作安排。
Step 3: 希望上級准假。

實際E-mail範例

寫信 ▼	刪除	回覆 ▼	寄件者： Ginny

Dear Sir/Madam,

Introduction

I am writing this letter to request a six-month parental leave, commencing on July 1st.

Body

I am expecting my first baby on June 6th, and plan to take the parental leave after my maternity leave. I have attached my applications for both maternity leave and parental leave.

I have talked to Ms. Julie Lai and she is willing to act for me during my leave.

Closing

I hope you will consider my request and grant me leave.

If any further information is required, I can be reached at extension 1234.

Thank you very much.

Best regards,
Signature of Sender
Sender's Name Printed

E-mail中譯　　寫信▼　刪除　回覆▼　　寄件者：Ginny

您好：

....

開頭

我寫這封信來請您准予我六個月的育嬰假，從 7 月 1 日開始。

本體

我的第一個小孩預計在 6 月 6 日出生，因此我計劃在產假之後直接放育嬰假。隨信附上產假及育嬰假的申請單。

我已經跟賴朱莉小姐談過，她願意在我休假期間幫我代理職務。

結尾

希望您能考慮我的請求，並准予我休假。

如果需要更多資訊，撥打分機 1234 就可以聯絡到我。

非常感謝您。

....

敬祝安康，
簽名檔／署名

各段落超實用句型

說明：畫底線部分的單字可按照個人情況自行替換

開頭 Introduction

❶ **I am writing this letter to request for a <u>30-day</u> leave of absence.**

我寫信來請您給我<u>三十天</u>的休假。

❷ **The purpose of this letter is to request for <u>a leave of absence</u> for <u>five days</u> starting from <u>May 25th</u>.**

我寫這封信目的是希望能從 <u>5 月 25 日</u>開始，請<u>五天事假</u>。

❸ **With this letter I would like to request for a <u>maternity leave</u> for <u>three months</u>, commencing on <u>July 1st</u>.**

藉由這封信，我想從 <u>7 月 1 日</u>開始，請<u>三個月育嬰假</u>。

❹ I am planning to take my **annual leave** from December 1st to December 18th, before **the Christmas holidays**.

我計劃在聖誕假期之前，從 12 月 1 日至 12 月 18 日休年假。

❺ I am writing to ask for a **one-month paternity leave**, starting from October 1st.

我寫這封信是要要求一個月的父親育嬰假，從 10 月 1 日開始。

本體 Body ······

❶ I have discussed with my husband and decided that we cannot afford a nanny.

我已經跟先生討論過，認為我們無法負擔請保姆的費用。

❷ I am using my annual leave to take a trip to **New Zealand and visit my grandparents**.

我將利用年假到紐西蘭去探望我的祖父母。

❸ Attached is my leave application.

附件為我的休假申請書。

❹ **Murray** will cover my shifts during my leave of absence.

穆瑞會在我休假期間幫我代班。

❺ **Jennifer** will act for me while I am away.

珍妮佛在我休假期間會代理我。

❻ I am going to need someone to cover my position while I am on leave, and **Jenny** is willing to help me out.

我休假期間將需要有人代替我的職務，而珍妮願意幫我的忙。

❼ While I understand that this is a busy time for our business and it may cause some inconveniences, the baby simply can't wait to pop out!

我瞭解我們公司這段時間相當繁忙，請假也許會造成不便，但寶寶就是快要生出來了！

結尾 Closing

❶ I hope you will consider my request and grant me leave.
我希望您能考慮我的申請，並准予我休假。

❷ If any further information is required, I can be reached by email or my extension number 1234.
若需要任何進一步的資訊，請以電子郵件聯絡我，或是撥打我的分機 1234。

❸ Please let me know whether this application is approved or not.
無論申請是否核准，都請您通知我。

❹ Thank you for approving my leave application.
感謝您核准我的休假申請。

❺ Please kindly consider my request.
請您考慮我的請假請求。

❻ If you would need any further information regarding this matter, please don't hesitate to contact me.
若您需要關於此事的更多資訊，請不吝與我聯絡。

 不可不知的實用E-mail字彙

- **leave of absence** 休假
- **leave application** 休假申請書
- **grant sb.'s leave** 准假
- **request for leave** 申請休假
- **leave without pay** 留職停薪

03 回覆抱怨與客訴

A Customer Complaints Reply Letter

 實際E-mail範例

寫信 | ▼　　刪除　　回覆 | ▼　　　寄件者： Ginny

Dear Sir/Madam,

Introduction

It is with great regret to hear that you were unsatisfied with the service you received at our Xin Yi Branch Restaurant last Sunday.

Body

We hope you will accept our sincere apologies for the unpleasant experience you and your family had at our restaurant.

Closing

We are giving our staff special training to strengthen their customer service skills. We assure you that such an incident will not occur again in the future.

Best regards,
Signature of Sender
Sender's Name Printed

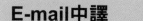

E-mail中譯　[寫信｜▼]　[刪除]　[回覆｜▼]　寄件者：[Ginny]

您好：

開頭

很遺憾得知您上週日對我們信義分店所提供的服務不滿意。

本體

對於您與家人在我們餐廳不愉快的經驗，我們希望您能接受我們所致上的誠摯歉意。

結尾

我們已經給予員工特別訓練，以加強他們的顧客服務技能。我們向您保證，這樣的事情不會再發生。

敬祝安康，
簽名檔／署名

各段落超實用句型

說明：畫底線部分的單字可按照個人情況自行替換

開頭 Introduction

❶ **In reply to your complaints regarding our after-sales service, ...**
在此回覆您對於本公司售後服務的抱怨，……

❷ **We regret to hear that you are not satisfied with the service you received at our store.**
很遺憾聽到您對您在我們餐廳得到的服務不滿意的消息。

❸ **I have received your email complaining about the quality of our products.**
我已經收到您抱怨本公司產品品質的來信。

❹ **I am writing this letter in answer to your complaints about our delivery service.**
我寫這封信是要回覆您對本公司配送服務的抱怨。

❺ I am regretful to hear that you found our service unsatisfactory.
很遺憾聽到您認為我們的服務令人不滿意。

❻ We are deeply sorry to hear that you have had an unsatisfactory experience with our services.
聽到您對於我們的服務有不太愉快的經驗，我們非常遺憾。

本體 Body

❶ I would like to apologize for your disturbing experience in our store.
我要為您在我們店裡得到的不愉快經驗向您道歉。

❷ I would like to express my deepest apologies for our unsatisfactory service.
我要為我們不符要求的服務向您至上最深的歉意。

❸ We hope you will accept our sincere apologies.
我們希望您能接受我們誠摯的歉意。

❹ Please accept our apologies for our staff's inappropriate service attitude.
請接受我們對員工不恰當的服務態度向您致歉。

❺ I would like to extend my apologies to you for the substandard service you received in our restaurant.
我想為您在本公司餐廳得到的不合格服務，向您致歉。

❻ We will look into the matter right away and deal out any punishments necessary.
我們將立即瞭解這件事，並進行該有的懲處。

結尾 Closing

❶ We are giving our staff special training on their customer service skills.
我們正針對顧客服務技巧，給予員工特別訓練。

❷ The waiter in question has been dismissed from his job.
有問題的服務生已經被解僱了。

❸ We assure you this kind of incident will never happen again.
我們向您保證，這樣的事情將不會再發生。

❹ We will exert our efforts to strengthen our <u>after-sales</u> service.
我們會盡一切努力加強售後服務。

❺ We will evaluate and improve our customer policies in order to make sure our staff is always prepared to provide satisfactory service to our customers.
我們將檢討並改進我們的客戶方針，以確保我們同仁永遠準備好為顧客提供滿意的服務。

❻ We have attached <u>a voucher to free spa services in our resort</u>, valid <u>the next three months.</u>
我們附上了本度假村的免費 SPA 服務券，接下來三個月都可使用。

 不可不知的實用E-mail字彙

- **after-sales service** 售後服務
- **customer service** 顧客服務
- **delivery service** 配送服務
- **service attitude** 服務態度
- **unsatisfactory service** 令人不滿的服務
- **substandard quality** 不符要求的品質
- **strengthen customer service skills** 加強客服技巧
- **special training on customer service** 客戶服務特訓

04 提出離職要求

A Resignation Request Letter

 實際E-mail範例

寫信｜▼	刪除	回覆｜▼	寄件者： Ginny

Dear Sir/Madam,

Introduction

It is with great regret that I have to notify you by means of this letter that I am resigning from my position as an assistant editor, effective March 1st.

Body

The reason of my resignation is that I want to switch professions.

Closing

Thank you for your support and guidance during the past two years.

It has been a pleasure working for you and with my colleagues at JBS Company.

I wish you continued success and all the best in the future.

Best regards,
Signature of Sender
Sender's Name Printed

E-mail中譯

寫信 ▼　　刪除　　回覆 ▼　　寄件者： Ginny

您好：

開頭

很遺憾以這封信通知您，我將於 3 月 1 日起，辭去我助理編輯的職務。

本體

離職的原因是因為我希望能轉換工作跑道。

結尾

感謝您過去兩年來的支持與指導。

能夠為您工作以及與 JBS 公司的同事們共事是我的榮幸。

祝您事業成功，並且未來一切順利。

敬祝安康，
簽名檔／署名

各段落超實用句型

說明：畫底線部分的單字可按照個人情況自行替換

開頭 Introduction

❶ I would like to inform you that I am resigning from my position as Chief Editor, effective May 21st.

我想通知您我將從 5 月 21 日起辭去我總編輯的職務。

❷ I am writing to submit my official resignation from JBS Company, effective one month from today.

我寫信來提交自今天起一個月後離開 JBS 公司的正式辭職信。

❸ It is with regret that I have to inform you that I wish to resign from my position, effective next month.

很遺憾通知您，我希望於下個月離職。

❹ I am sorry to inform you by this letter that I have decided to resign.

很抱歉以此信通知您，我已經決定要離職了。

❺ I am regretful that I have decided to quit.
很抱歉我已經決定要離職了。

❻ I am writing this email to inform you that due to <u>personal reasons</u>, I have decided to resign from my position.
我寫這封信是要告知您，由於<u>個人因素</u>，我已決定離職。

本體 Body ....

❶ My last day at work will be <u>February 29th</u>.
<u>2 月 29 日</u>將會是我最後一天上班。

❷ I will remain available at my place of work till the last day of my work.
在我離職之前，我都會一直在我的崗位上服務。

❸ The past <u>two years</u> at <u>JBS Company</u> have been really rewarding.
在 <u>JBS</u> 公司工作的過去<u>兩年</u>，真的很使我獲益良多。

❹ I would like to take this opportunity to thank you for your support and instruction during the past <u>two years</u>.
我想藉此機會感謝您過去<u>兩年</u>來的支持與指導。

❺ I make this decision because I want to <u>switch professions</u>.
我做這個決定，是因為我想要<u>轉換事業跑道</u>。

❻ I have had a really great time working here, and I learned a lot from everyone in the company.
我在這裡工作得非常愉快，和公司裡的大家學到了很多。

結尾 Closing ....

❶ It's been a pleasure working for you and with all my colleagues.
為您工作，以及與所有同事共事，是我的榮幸。

❷ I appreciate your professional guidance. I have learned a lot from you.
感謝您的專業指導。我從您身上學到很多東西。

❸ I will be pleased to help training my substitute during the transition.
我會很樂意在交接期間協助訓練接替我的人。

❹ I am willing to offer my assistance in the transition of responsibilities.
我很願意在職務交接時提供協助。

❺ I wish you and **JBS Company** continued success.
祝您與 JBS 公司持續成功。

❻ Working here at **JBS Company** will be an experience that I will forever cherish.
在 JBS 公司工作的這段時間，將會是我永遠珍惜的經驗。

 不可不知的實用E-mail字彙

- **transition of responsibilities** 職務交接
- **train substitute** 訓練接替的人
- **professional guidance** 專業指導
- **switch professions** 轉換工作跑道
- **the last day of work** 最後一天上班
- **official resignation** 正式辭職
- **resignation from sb.'s position** 辭去職務；離職

05 推薦合適人選

A Recommendation Letter

寫作要點 Key Points:

Step 1: 表示來信目的為推薦適合人選。
Step 2: 具體說明推薦該人選之原因。
Step 3: 感謝對方予以考慮。

實際E-mail範例

| 寫信 ▼ | 刪除 | 回覆 ▼ | 寄件者： Ginny |

Dear Sir/Madam,

Introduction

It is with great pleasure that I am writing this letter of recommendation for Ms. Becky Huang. I am very glad to recommend her for the position of Chief Editor.

Body

I have been working closely with Becky for three years and have always been impressed by her ability to complete all her jobs and at the same time offer high quality of work on time.

Closing

I am very confident that Becky is the right person for the position of Chief Editor.

I appreciate your consideration.

Best regards,
Signature of Sender
Sender's Name Printed

E-mail中譯

寫信 ▼　刪除　回覆 ▼　寄件者：Ginny

您好：

開頭

很高興為黃佩琪小姐寫這封推薦信。

我很樂意推薦她擔任總編輯一職。

本體

我已經與佩琪近距離工作三年的時間，並且一直很欣賞她能夠準時完成交付工作，且同時兼顧工作品質的能力。

結尾

我相信佩琪會是擔任總編輯的不二人選。

感謝您考慮我的推薦。

敬祝安康，
簽名檔／署名

各段落超實用句型

說明：畫底線部分的單字可按照個人情況自行替換

開頭 Introduction

❶ **I am very glad to have the opportunity to recommend Susan Gales.**
我很高興有機會向您推薦蘇珊凱爾。

❷ **It is my pleasure to recommend Alexander for the position of Chief Editor with ABC Book Publisher.**
能推薦亞歷山大擔任 ABC 圖書出版公司的總編一職，是我的榮幸。

❸ **I am very pleased to recommend Anna for the position of Sales Manager with JBS Company.**
我很高興能向 JBS 公司推薦安娜擔任業務經理。

❹ It is with great pleasure that I am writing this recommendation for David Hampton.

很高興能為大衛漢普頓寫這封介紹信。

❺ With this letter I would like to recommend Ms. Kelly Hans to your company.

我想利用這封信，向貴公司推薦凱莉漢斯小姐。

本體 Body

❶ I have been very impressed by Steven's ability to complete the tasks assigned to him on time, or even ahead of time.

我一直很欣賞史帝文總是能準時，甚至提早完成所交付之工作的能力。

❷ He is an efficient and able employee.

他是個很有效率，而且很能幹的員工。

❸ He is a born leader. I believe your company will be even more prosperous under his leadership.

他是個天生的領袖。我相信貴公司在他的領導下會變得更加成功。

❹ Peter proved himself an excellent salesman as his sales performance consistently exceeded my expectation.

彼得的業績總是超過我的期待，證明自己是個有優秀的業務員。

❺ She is very creative and always contributes marvelous ideas during brainstorming sessions.

她非常有創意，總是能在集體研討會議上貢獻絕佳的好點子。

❻ He never fails to brighten up the office with his cheerful and positive personality.

他總是用樂觀正面的個性為辦公室帶來歡樂。

結尾 Closing

❶ I believe he will continue to find success in sales.

我相信他會在業務這個領域繼續不斷地成功。

❷ I have no doubt that he will lead your R&D Department to success.

我堅信他能帶領貴公司的研發部門走向成功。

❸ I look forward to seeing him in the position of the Marketing Manager.

我期待能看到他擔任行銷部經理一職。

❹ You will not regret having her as your HR Manager.

你絕不會後悔請她擔任您的人力資源經理。

❺ She is surely the right person for the position of Chief Editor.

她絕對是擔任總編輯的不二人選。

❻ I recommend her for the position with all my heart.

我全心推薦她擔任這個職位。

 不可不知的實用E-mail字彙

- **recommend** 推薦
- **letter of recommendation** 介紹信
- **the right person for the position of ...** 擔任～職務的不二人選
- **hard worker** 認真工作的人
- **efficient** 有效率的
- **complete the task assigned on time** 準時完成交付的工作
- **able employee** 能幹的員工
- **born leader** 天生的領袖

06 哀悼與慰問

An Condolence Letter

寫作要點 Key Points:

Step 1: 對對方所遭遇的事表示哀悼或慰問。
Step 2: 緬懷逝者。
Step 3: 為逝者祝願，並請對方節哀。

實際E-mail範例

   寄件者： Ginny

Dear Sir/Madam,

Introduction

With a heavy heart and sad feelings, I am writing this letter on behalf of JBS Company to offer our deepest sympathies on the sudden demise of Mr. Michael Jones.

Body

Mr. Jones was a true professional. His contribution to the growth of both of our companies can never be forgotten. His death is a huge loss to us.

Closing

We pray to God that his soul rest in peace.

Our sincere thoughts are with you.

Best regards,
Signature of Sender
Sender's Name Printed

E-mail中譯

| 寫信▼ | 刪除 | 回覆▼ | 寄件者： Ginny |

您好：

開頭

忍著沈重的心情與悲傷的情緒，我謹代表 JBS 公司寫這封信，表達我們對麥可瓊斯先生突然過世一事的最深哀悼。

本體

瓊斯先生是個真正的專業人士，他為我們兩家公司的成長所作的貢獻，將永存我們心中。他的過世對我們來說是個極大的損失。

結尾

我們向上帝祈禱，願他的靈魂獲得安息。

我們誠摯的心與您同在。

敬祝安康，
簽名檔／署名

各段落超實用句型

說明：畫底線部分的單字可按照個人情況自行替換

開頭 Introduction

❶ We are very sad to hear about the death of Mr. Simon Blues.

得知賽門布魯斯過世的消息讓我們十分難過。

❷ Please accept my deepest sympathies for the loss of Jeffery.

請接受我對失去傑佛瑞最深的慰問。

❸ We would like to convey our deepest condolence on the death of Mr. Cole.

我們想表達我們對科爾先生過世最深的哀悼之意。

❹ It is with a heavy heart and sad feelings that I am writing this letter on behalf of JBS Company.

帶著沉重的心與悲傷的情緒，我代表 JBS 公司寫這封弔唁信。

❺ On behalf of JBS Company, I would like to convey our sympathies to the family and friends of Ms. Jennifer Woods.

謹代表 JBS 公司，我在這裡表達對珍妮佛伍茲小姐的家人與朋友的弔唁問候。

❻ I am writing this email to express my deepest condolences for the loss of Mr. Wang.

我寫這封信，是要對於王先生離世表示深深的遺憾。

本體 Body

❶ My heart is full of sorrow.

我的心充滿悲傷。

❷ He was a good caring friend as well as a great client.

他是個有愛心的好朋友，也是個很棒的客戶。

❸ He was a hardworking and trustworthy employee.

他是個努力工作並且值得信賴的員工。

❹ We will always remember him in our thoughts.

我們將永遠在心裡緬懷他。

❺ His contribution to the growth of the company will always be remembered.

他對公司的成長所作的貢獻將永遠會被記住。

❻ I will always keep his kind words and gentle manners in my heart.

他和善的話語與溫和的態度，我將永遠銘記在心。

結尾 Closing

❶ My thoughts are with you during the time of bereavement.

在這段治喪期間，我的心思將與你們同在。

❷ Please do not hesitate to contact us whenever you need assistance. We are always ready to help you.

無論何時只要您需要協助，請儘管與我們聯絡。我們隨時都準備好幫助您。

❸ If there is anything you need during this difficult time, please do not hesitate to let us know.

在這段艱苦的時候，有任何需要請儘管讓我們知道。

❹ He will be greatly missed by all of us.

我們將會非常懷念他。

❺ We pray that God will give you strength to bear this great loss.

我們向上帝祈求，希望祂能賜予你們承受失去摯愛的力量。

❻ I understand that this must be a difficult time for you. Please do not hesitate to let me know if there's anything you need.

我瞭解這段時間對您而言一定不好過。若您有需要，請不吝與我聯絡。

 不可不知的實用E-mail字彙

- condolence　慰問，哀悼
- condolence letter　弔唁信
- convey sympathies　表達慰問之意
- sudden demise　驟逝
- the death of sb　某人之死
- bereavement　失喪；喪親（友）之痛

07 對外澄清誤會

A Misunderstanding Clear-up Letter

寫作要點 Key Points:

Step 1: 說明來信目的為澄清某個誤會。
Step 2: 簡短而清楚地重述相關內容。
Step 3: 表達歉意，並請對方諒解。

實際E-mail範例

| 寫信 ▼ | 刪除 | 回覆 ▼ | 寄件者： | Ginny |

Dear Sir/Madam,

Introduction

I have just realized that you have a misunderstanding about the shipping date regarding our order for your computer components.

Body

I hereby would like to confirm with you again that we hope to receive our order by October 31st. That means you need to have our order dispatched ahead of time.

Closing

I should have made myself clear on what I said.

Please accept my apology.

Best regards,
Signature of Sender
Sender's Name Printed

E-mail中譯　　寫信▼　刪除　回覆▼　寄件者：Ginny

您好：

....

開頭

我剛剛才明白，原來您對我們向您訂購電腦零件的訂單之交貨日期有所誤解。

本體

我在此再跟您確認一次，我們希望能在 10 月 31 日之前收到我們的訂單。也就是說，您必須提前派送我們的訂單。

結尾

我應該要把話說清楚一點才對。

請接受我的道歉。

....

敬祝安康，
簽名檔／署名

各段落超實用句型

說明：畫底線部分的單字可按照個人情況自行替換

開頭 Introduction

❶ **I just realized that we have a misunderstanding about my opinion toward your article.**

我剛剛才知道我們在我對您文章的看法上有所誤會。

❷ **I am writing this letter to clear up a misunderstanding between us.**

我寫這封信是要釐清我們之間的誤會。

❸ **I think there is a bit of a misunderstanding between us regarding the job assignments.**

我想我們在工作分配方面有點小誤會。

❹ I am afraid that you have misunderstood what I said.
你恐怕誤解了我所說的話。

❺ I regret that my words have been totally misinterpreted.
很遺憾我所說的話被完全地誤解了。

❻ I just learned that there has been a bit of a misunderstanding, likely due to my confusing wording in my most recent email.
我剛剛才知道發生了一點誤會，很可能是我在最近一封電子郵件中用詞有疑慮造成的。

本體 Body

❶ This kind of misunderstanding could have been avoided if I were more careful with my words.
如果我多注意我的用字遣詞，就可以避免這樣的誤會了。

❷ I hereby would like to explain our return and refund policies to you again.
我希望能在此再跟您解釋一次本公司的退換貨政策。

❸ I hereby would like to reaffirm that we do not have any branch restaurants at the moment.
我要在此重申，我們餐廳目前並沒有任何分店。

❹ Allow me to reaffirm here that our company does not have a partnership with POP Company.
容我在此重申，本公司跟 POP 公司並無任何合股關係。

❺ I must emphasize again that I did not reveal any confidential information regarding the project to the third party.
我必須再次強調，我並沒有向第三方透露任何有關此專案的機密資料。

❻ What I meant in the email was that the policy will be effective starting the next year, not that it will be effective until the next year.
我在電子郵件中的意思是，這個政策明年將開始生效，而不是生效日期到明年為止。

結尾 Closing

❶ **I apologize for causing you inconvenience.**
造成您的不便，在此向您致歉。

❷ **Please accept my apology if I hurt your feelings.**
如果我讓你不舒服，請接受我的道歉。

❸ **I should have been more cautious with my words.**
我的用字遣詞應該要更謹慎才是。

❹ **I am regretful for contributing to this misunderstanding.**
很抱歉造成這樣的誤會。

❺ **I apologize again for all the trouble that this misunderstanding has caused.**
再次對這件誤會所造成的所有困擾，向您道歉。

❻ **I am very sorry about my terrible wording and assure you that I will proofread my emails more carefully in the future.**
我很抱歉用詞不更清楚，且保證未來在寫電子郵件時將更加小心。

 不可不知的實用E-mail字彙

- **misunderstanding** 誤會
- **contribute to a misunderstanding** 造成誤會
- **explanation** 解釋
- **clear up a misunderstanding** 釐清誤會
- **misinterpretation** 誤解

An Office Supplies/ Stationary Request Letter

寫作要點 Key Points:

Step 1: 說明來信目的為申請辦公用品或公司內部物品
Step 2: 具體表示申請數量。如必要需附上申請單。
Step 3: 感謝對方協助申請。

實際E-mail範例

寫信 ▼　刪除　回覆 ▼　　寄件者： Ginny

Dear Sir/Madam,

Introduction

I am writing this letter to inform you that we are running short on supplies and need your assistance in ordering them.

Body

The following is the list of supplies needed in the office:

1. paper clips

2. binder clips

3. staples

4. black printer cartridges

Closing

We hope to receive the above items before the week ends.

Thank you for your assistance.

Best regards,
Signature of Sender
Sender's Name Printed

E-mail中譯　　寫信│▼　　刪除　　回覆│▼　　寄件者：Ginny

您好：

....

開頭

我想通知您，我們有些辦公室用品已經快用完了，需要您協助訂購。

本體

以下為辦公室所需要的用品清單：

1. 迴紋針

2. 長尾夾

3. 釘書針

4. 黑色碳粉匣

結尾

我們希望能在這週結束之前收到以上物品。

感謝您的協助。

....

敬祝安康，
簽名檔／署名

各段落超實用句型

說明：畫底線部分的單字可按照個人情況自行替換

開頭 Introduction

❶ I am writing this letter to request an <u>electric shredder</u> for our department.

我寫這封信是要為我們部門申請一台<u>電動碎紙機</u>。

❷ We would like to request approval to purchase a <u>laser printer</u> for <u>Research & Development</u> Department.

我們想申請同意為<u>研發部</u>門購買一台<u>雷射印表機</u>。

❸ I would like to inform you that we are requesting for supplies that we need in the office.

我想通知您，我們想申請辦公室所需的用品。

❹ I would like to request for the supply of stationery items which we are currently short of.

我想申請我們目前短缺的文具用品。

❺ I am writing to request the following stationery items for daily use in the office.

我是寫信來申請以下每日所需的辦公室文具用品。

本體 Body

❶ Attached is the stationery item request form.

附件為文具用品申請單。

❷ The following items are urgently needed for the daily activities in R&D Department.

研發部門目前急需以下物品，以供日常辦公活動所需。

❸ The items we need the most at the moment are staples, L file holders and A4 copy paper.

我們目前最需要的物品是釘書針、L 夾和 A4 影印紙。

❹ We will be printing a big project so we are in urgent need of A4 copy paper.

我們將要列印一項大項目，所以我們急需 A4 影印紙。

❺ We are running out of both black and color printer cartridges.

我們的黑色及彩色印表機碳粉匣都用完了。

結尾 Closing

❶ Looking forward to your timely response.

期望您盡快回覆。

❷ Thank you for your assistance.

感謝您的協助。

❸ Please contact Sarah at #123 if additional information is needed.

如果需要其他資訊，請撥打分機 123，與莎拉聯絡。

❹ I hope we will receive these items before <u>the week ends</u>.
　希望這週結束前可以收到這些物品。

❺ I hope you grant our request. Thanks.
　我希望您能同意我們的請求。謝謝。

 # 不可不知的實用E-mail字彙

- **correction tape**　立可帶
- **scissors**　剪刀
- **paperclip**　迴紋針
- **binder clip**　長尾夾
- **file folder**　資料文件夾
- **L-file holder**　L夾
- **post-it**　備忘貼
- **calculator**　計算機
- **printer**　印表機
- **copier**　影印機
- **shredder**　碎紙機
- **stapler**　釘書機
- **staple**　釘書針
- **printer cartridges**　印表機碳粉匣

原來如此 系列 *E167*

一本戰勝！商務英文E-mail全情境指南，商務英文書信抄這本就搞定！

商務英文書信抄這本就搞定！

作　　者	蔡文宜	
顧　　問	曾文旭	
社　　長	王毓芳	
編輯統籌	耿文國、黃璽宇	
主　　編	吳靜宜、姜怡安	
執行主編	李念茨	
執行編輯	陳儀蓁	
美術編輯	王桂芳、張嘉容	
封面設計	阿作	
法律顧問	北辰著作權事務所　蕭雄淋律師、幸秋妙律師	

初　　版	2017年07月初版1刷 2019年再版2刷
出　　版	捷徑文化出版事業有限公司
電　　話	（02）2752-5618
傳　　真	（02）2752-5619
地　　址	106 台北市大安區忠孝東路四段250號11樓-1

定　　價	新台幣350元／港幣117元
產品內容	1書

總 經 銷	采舍國際有限公司
地　　址	235 新北市中和區中山路二段366巷10號3樓
電　　話	（02）8245-8786
傳　　真	（02）8245-8718

港澳地區總經銷	和平圖書有限公司
地　　址	香港柴灣嘉業街12號百樂門大廈17樓
電　　話	（852）2804-6687
傳　　真	（852）2804-6409

本書圖片由Shutterstock提供

捷徑 Book站

現在就上臉書（FACEBOOK）「捷徑BOOK站」並按讚加入粉絲團，
就可享每月不定期新書資訊和粉絲專享小禮物喔！

http://www.facebook.com/royalroadbooks
讀者來函：royalroadbooks@gmail.com

國家圖書館出版品預行編目資料

一本戰勝！商務英文E-mail全情境指南，商務英
文書信抄這本就搞定！/ 蔡文宜著. -- 初版. -- 臺北
市：捷徑文化, 2017.07
　　面；　公分（原來如此：E167）
ISBN 978-986-94544-7-6(平裝)

1.商業書信 2.商業英文 3.商業應用文 4.電子郵件

493.6 　　　　　　　　　　　106008101